S M O K E

Cigars, Cigarettes, Pipes, and Other Combustibles

K.M. Kuntz

TODTRI

ACKNOWLEDGMENTS

Todtri Productions would like to thank the following individuals and organizations for their assistance: Cigar Aficionado, Colibri, Cuba Club, David W. Dyson, Jacques Halbert, High Times Magazine, Kim McCarty, Modernism Gallery, Museum of The American Indian, Antonio M. Rosario, and Jerry Striker.

Copyright © 1997 by Todtri Productions Limited. All rights reserved. No part of this publication may be reproduced, stored in a retrieval system or transmitted in any form by any means electronic, mechanical, photocopying or otherwise, without first obtaining written permission of the copyright owner.

This book was designed and produced by Todtri Productions Limited P.O. Box 572, New York, NY 10116-0572 FAX: (212) 695-6988

Printed and bound in Korea

ISBN 1-57717-062-8

Author: K. M. Kuntz

Publisher: Robert M. Tod
Editorial Director: Elizabeth Loonan
Senior Editor: Cynthia Sternau
Project Editor: Ann Kirby
Photo Editor: Meiers Tambeau
Production Coordinator: Jay Weiser
Designer: Theresa Izzillo

PHOTO CREDITS

Art Resource, New York 44
   Bridgeman 21 (top)
   Giraudon 20, 74–75
   Scala 35
Cigar Aficionado 59, 64 (bottom), 66 (bottom)
Colibri 123 (right)
Corbis-Bettmann, New York 4, 10, 11, 13, 15, 16 (top), 22–23, 24, 26, 27, 29, 30, 32 (bottom), 38, 39 (top), 46, 54 (bottom), 55, 82, 85 (right), 86, 87 (top & bottom), 90, 92 (left), 92–93, 96 (left), 96–97, 107, 110, 111 (left), 113, 119 (bottom), 120–121
Cuba Club, Miami 76
Culver Pictures, Inc. 64 (top)
David W. Dyson 9, 16 (bottom), 109
Gamma Liaison Network
   Jeff Katz 126 (top)
Globe Photos 99, 127
   Steve Finn 65
   Sonia Moskowitz 112
Jacques Halbert 61
Rosemarie Hausherr 28 (bottom)
High Times Magazine 104
Henry Horenstein 79, 103
International Press Online, Inc. 17
Jewish Museum/Art Resource, New York 119 (top)
Kobal Collection 6, 25, 43, 60, 62–63, 83, 89, 94, 100, 114, 116–117, 118
   Linda R. Chen 84–85
   Lorey Sebastian 122–123
London Features
   David Fisher 91 (top)
   D. Ridgers 95 (top)
Kim McCarty 33
James R. Minchin III 7, 41
Modernism Gallery 71 (top)

Museum of The American Indian 14 (top & bottom)
Omni-Photo
   Joan Slatkin 98
Patrick Pagnano 68–69
Photofest 81
Photophile 31, 57
Picture Perfect
   Sharon Cummings 47
Private Collection 12, 18, 54 (top), 56 (top), 88 (top), 91 (bottom), 102
Marisa Pryor 70
Carl Purcell 36–37, 42
H. Armstrong Roberts 49, 58
The Norman Rockwell Museum (printed by permission of the Estate of Norman Rockwell, © 1960) 50
Antonio M. Rosario 19, 78, 126 (bottom)
Shooting Star 5, 45, 52–53, 115 (top)
   Yoram Kahana 105
   Donald Smetzer 67
Jerry Striker 28 (top), 32 (top), 34, 56 (bottom), 66 (top), 72 (left & right), 77, 115 (bottom)
January Stuart 80
Tate Gallery, London/Art Resource New York 106
John Tenniel 111 (right)
Touchstone/Shooting Star 124–125
Transparencies Inc.
   J. G. Faircloth 108
Underwood Photo Archive 40, 88 (bottom)
Unicorn Stock Photos
   Betts Anderson 116 (left)
   Arni Katz 48
   Jim Shippee 101
   Visual Contact 51, 73
Lorraine Wardy Enterprises
   Fernando Pena 71 (bottom) Michael Worthy 21 (bottom), 39 (bottom)

CONTENTS

INTRODUCTION

4

Chapter One
TOBACCO THROUGH THE AGES

10

Chapter Two
THE PIPE

34

Chapter Three
THE CIGAR

58

Chapter Four
THE CIGARETTE

82

Chapter Five
THE TOBACCO MYSTIQUE

106

INDEX

128

"The greatest service that can be rendered any country is to add a useful plant to its culture."

—Thomas Jefferson

Introduction

What can make one person swoon and another grimace in disgust, turn noses up and away or lift spirits to an ethereal place, and definitively separate human beings from all other species on (and probably off) the planet? Tobacco. Tobacco and smoking have continually held their ground in the feistiest of arenas, luring newcomers to praise or condemn the most widely recognized social habit known to humankind, and spawned—in the meantime—an international industry of colossal proportion.

Despite what some will say, there is a degree of beauty in all of this and a fascinating story to be told. Tobacco is a bit like the fly on the wall we've all hoped to be: This unremarkable green leaf has been everywhere across time and space and, until recently, has enjoyed an open door at even the most private of gatherings. Tobacco *is* history, society, style, and human nature all rolled (literally) into one. To trace the origin of tobacco is to track a footpath into infinite realms of experience. If we could glimpse where its smoke has been, we would indeed be privy to a vast amount of knowledge.

The opportunity to write this book came at a perfect time. I'd just returned from living in Thailand for two years and had a clean slate on which to work. Uncertain of the angle I would take and knowing the breadth of the topic, I approached it with very much the same attitude I have regarding it: open-minded curiosity. What I found was a veritable universe of information, insight, and passion—both in favor of tobacco and against it—upon which to set my sights.

It seemed every time I attempted to narrow the scope of my subject matter, by opening a book or accessing a website, I stumbled upon another fascinating piece of history, another tale to tell, another valid anti-smoking perspective, another twist of fate, another book to

Taking sophistication to the next level: Marlene Dietrich donned men's suits and sported *the* men's accessory in the 1930s, creating a new, tough image of femininity.

INTRODUCTION

write. It occurred to me then, that I had accepted far too large a project; suddenly I realized that I'd been devilishly tricked into recording the history of the world. Tobacco has, perhaps more so than any other product been with us through seemingly everything.

I was very lucky to be in New York City and have access to the Public Library. In that intimidating edifice rests the greatest stories ever untold. It was my pleasure to dig through The Arents Collection in the rare books and manuscripts section. Anyone familiar with tobacco—in all its forms and various manifestations throughout history—should know the name George Arents Jr. Mr. Arents made his fortune in tobacco as a co-founder of the American Machine & Foundry Company (AMF), one of the first makers of cigar and cigarette-rolling machinery. He was, perhaps, the greatest lover and most dedicated collector of any literature that contained even the smallest reference to the herb. At the time of his death in 1960, his library contained more than twelve thousand rare books and manuscripts in almost thirty foreign languages. His collection also houses various works of art and music that deal with tobacco, and hundreds of periodicals, cigarette cards, cigar boxes and bands, snuff boxes, pipes, and more. Subsequent to his death, Jerome Brooks and Sarah A. Dickson translated and catalogued the contents of Arents' vast library in five volumes loaded with history and innumerable tidbits for smoky conversation fodder.

He fancied first and rare editions signed by the author, if possible, and went to great lengths to obtain these. His collection contains many unique pieces from both the sixteenth and seventeenth centuries, and reaches into the eighteenth century and early nineteenth centuries as well. The worn pages of these well-loved books and Arents' own handwritten notes alone

Right: **Smoking was the least of Uma Thurman's vices in the 1994 instant cult classic** *Pulp Fiction.*

Opposite: **Modern-day blues master Taj Mahal takes to the road with the elements of nineties style—shades and a fine cigar.**

made my own research worthwhile. By the twentieth century, as the popularity of smoking grew in both use and image, the material available to Arents became too much and forced him to narrow his attention, thereby forfeiting a portion of the burgeoning tobacco culture. Luckily, many connoisseurs and authors have generously continued to develop this outstanding collection on tobacco and smoking, hence contributing to its continuing tradition as one of the greatest personal libraries in the world and the foremost comprehensive collection on tobacco-related material.

To those writers and lovers of their smoke, I owe a great deal of thanks for both the wealth of information they provided me and the joy they so obviously derived from the presentation. By far the finest modern-day book on pipes and pipe smoking is *The Ultimate Pipe Book*, by Richard Carleton Hacker. His knowledge of and love for the pipe come through in every sentence of his book. Smoking czars, Zino Davidoff and Alfred Dunhill, also wrote fascinating tributes regarding the significance of tobacco throughout the ages.

What I've put down in this book merely skates over a thin top layer of the full story. We only know of the Native Americans usage of tobacco through the incomplete and somewhat biased accounts of European explorers. Women must have a fascinating smoking history all their own, but the thought of women smoking was so scandalous for so long that only a fraction of isolated incidents have been recorded and repeated here. Surely there was more to this centuries-old

ritual than what the colonists saw. The past has undoubtedly more than the few famous smokers mentioned here, and they could have certainly offered me different insights and anecdotes into the trends of their time. Furthermore, the international smoking scene is a burgeoning one (as are global and widely supported anti-smoking campaigns), and provides a great confirmation of the powers of both tobacco the plant and tobacco the industry. Alas, I was confined to only one volume and much had to be left behind for another's research. I hope I have succeeded in offering a broad and entertaining scope of the subject.

This book, then, is a chronicle of tobacco's journey to the forefront of our society and the center of our debates. I've attempted to show how tobacco has courted its lovers and

Introduction

Always debonair, Errol Flynn preferred a pipe to the more popular cigars and cigarettes of the day.

been glorified by the immense industry that produces its various products. There is some basic information on the centuries-old processes of growing, harvesting, and curing the leaves of the tobacco plant and some more on the corporations behind the products.

I've concentrated mostly on the numerous trends and transformations that tobacco has enjoyed throughout the past three hundred years and across various societies and continents. All the while, I've tried not to mask its well-known controversy with the attributes its users find appealing. Therefore, my goal is neither to glorify nor stigmatize this "bewitching herb," but rather to highlight the ways and reasons it's gained such notoriety and demonstrate how much attention is put into it from both its proponents and its opponents; to look at smoking and tobacco as a fascinating aspect of our culture and our history apart from the individual feelings it may engender.

After learning so much, I am also compelled to re-introduce an aspect of tobacco that seems to be lost in our great debate over its use. That is, how much this use has been altered from the original artform its founders practiced. In all my research, I've yet to find a proponent of chain-smoking mass produced and doctored forms of the herb. Rather, the true lovers of tobacco eschew its overuse, comparing it to a fine wine whose myriad flavors and sensations would be lost in a quick swill. Sadly, it seems we have forgotten that tobacco was originally smoked as an offering to gods during the rituals and ceremonies of Native Americans. Its use was sacred and its users had grave respect for its power.

As with any popular commodity, tobacco has been constantly re-shaped and re-marketed to fill the demands of societies obsessed with what's the newest and the finest. It's impossible not to be impressed, or at least surprised, when looking back at what this enormous industry has spawned. The wealth of gadgets and marketing campaigns presented by tobacco, pipe, cigar, and cigarette companies throughout history pales in comparison only to the depth of the public's desire for more. Smokers are privy to many of the most creative advertisements around. This book includes some of the more famous campaigns as well as pictures of some of the "paraphernalia" that's been developed.

Part of the allure of smoking has always been the result of high profile persons indulging in public. Some people like the comedian George Burns are so renown for their smoking that it is a primary part of the image conjured up by their name. Here you will find pictures of many of the world's most famous smokers along with quotes from many others. Both the familiar and rarely seen pictures are surely worth more than my few thousand words, so now, slip into a cloudy room and enjoy.

Dragon lighter, by David W. Dyson.

Back in the days when presidents drove their own cars, Franklin D. Roosevelt's cigarette holder was his trademark.

Poster for Job cigarettes, undated.

"Much smoking kills live men and cures dead swine."

—GEORGE D. PRENTICE

CHAPTER ONE

TOBACCO THROUGH THE AGES

A Chronology Of Tradition

History, written and spoken, shows us an unbroken habit of cultures using various plants and weeds for their intoxicant, medicinal, and social value. We know that the ancient Babylonians considered inhaling smoke through reed pipes to be both therapeutic and communal. Even Hypocrites, "the father of modern medicine," imbibed in an occasional toke. The Aboriginal peoples of Australia, whose societies date back well over twenty thousand years, regularly chewed the leaves of local herbs. Ancient Asians and Southern Europeans used betel-nut and bhang in similar ways as well.

Tobacco through the Ages

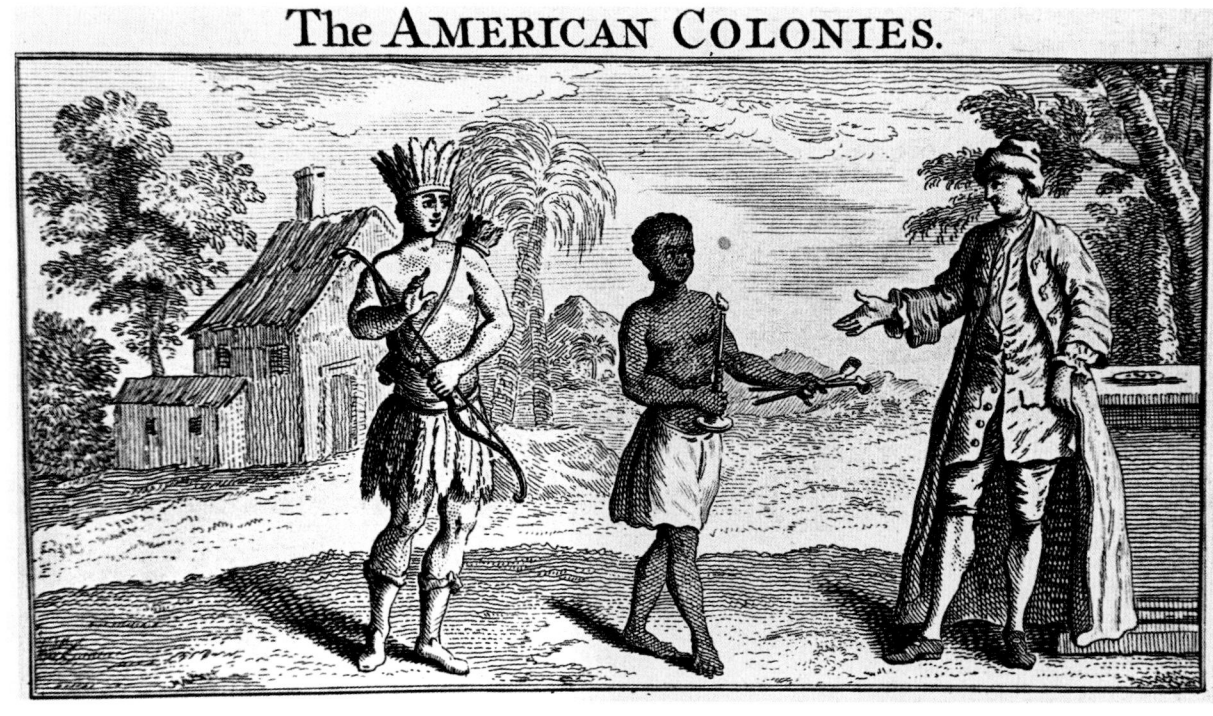

Fifteenth-century European explorers headed west for gold, God, and glory, and came back with an added bonus: tobacco.

The use of tobacco specifically though, has a history recorded much more recently. While its geographic origin is still a topic of debate, the general consensus is that tobacco came from the Americas and was used by the natives of those lands for thousands of years prior to its worldwide renown. Historians base this contention mostly on the complete absence of any reference to the plant in travel journals dated before those who explored the North and South American continents.

While some historians point to apparent species of the plant throughout Asia and Africa during this time as evidence to the contrary, it would seem an uncharacteristic oversight on the part of Western travelers to eliminate such a novel habit as smoking (then utterly unheard of in Europe) from their meticulously recorded observations. The extensive annals of Marco Polo, Vasco de Gama, and William de Rubruquis, who had traveled to the Asian continent before the 1500s, failed to make any mention of tobacco.

Tobacco through the Ages

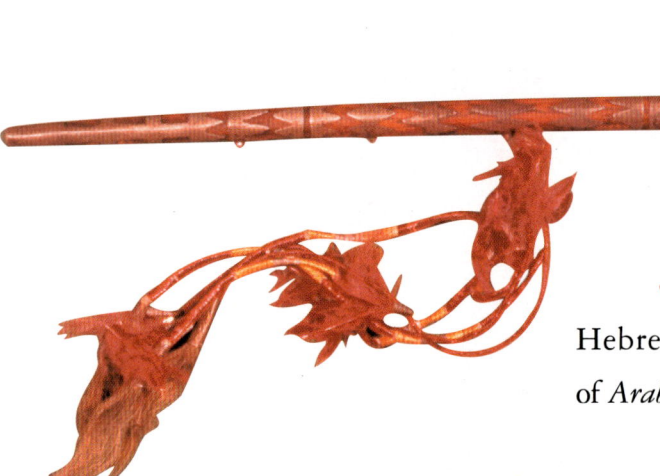

Right: The calumet, or peace pipe, was shared by Native Americans in religious ceremonies. Indians offered tobacco to Columbus when he landed at San Salvador, and he soon learned the value of the gift.

Above and below: Blackfoot Indian pipe tomahawks.

Also, there are no obvious references to tobacco in ancient Sanskrit manuscripts, Hebrew writings, throughout the text of *Arabian Knights*, or in the *Koran*.

Before Western Exploration Of The Americas

The Mayan civilization, which flourished almost two thousand years ago, depicted priests smoking out of pipes in their intricate stone carvings throughout southern Mexico. Knowing that tobacco was indigenous to this area, leads to the assumption that this society not only smoked tobacco, but may be responsible for spreading its use throughout Central and North America around the year 500. Thereafter, the sailors of that area continued to travel, spread the seeds, and established the use of this local weed well before Western peoples were introduced to the leaf.

Native inhabitants of the land and foreign visitors used tobacco in bartering and buying, and travel and trade. It may well be considered the first currency of the Americas and the perfect social buffer between and among hundreds of different tribes. An offer of tobacco was an offering of friendship and an opening of potential relations. Tobacco held a high position in both daily life and sacred ritual.

Native Americans had enormous respect for the weed and used it in religious ceremonies. The most famous image of tobacco use in these cultures is of the "peace pipe," or calumet, which smokers passed between one another as a sign of good will. Tobacco was burned over open flame to offer its sweet smell to the gods. It was used to anoint warriors and to placate angry spirits. After killing animals, hunters would send the victim's spirit up to heaven in a veil of smoke. Fisherfolk tossed tobacco to the sea as a request for calm waters. Most of these groups did not use it lightl;, smoking paraphernalia and sometimes tobacco itself were entrusted to a guardian and taken out only for special occasions.

The aristocracy of Pre-revolutionary France may have preferred cigarettes to cigars, but apparantly the revolutionaries knew—and Marie Antoinette learned—the value of a fine guillotine cut. *Right:* Scene from *Marie Antoinette,* 1938. *Below:* Guillotine cigar cutter by David W. Dyson.

Tobacco's Introduction To Western Civilization

Research credits Christopher Columbus with the first notations of tobacco in its use by the Arawak Indians of San Salvador in the Bahamas. His journals describe how these people greeted him with warmth and gave him a package of leaves. He observed how his hosts were smoking them and soon learned the social implications of their presentation to him. Shortly thereafter in around 1518, Grijalva and later Cortez, set out from Spain to explore the Yucatan Peninsula and Tabasco in Mexico. They saw the natives of that area use tobacco as Columbus had described, and also witnessed the artistry with which they cultivated the herb.

All of the explorers to the area came to understand this strange plant as an important and unifying aspect of the various cultures throughout these two continents. Unfortunately, most of the explorers also considered the natives of

SMOKE

Tobacco through the Ages

the land "heathens" and therefore saw the tobacco use as an "un-Christian" thing to do. Those that did partake in the ritual, did so for pleasure. The economic potential of anything from this "New World" did not cross anyone's mind for about another one hundred years, because the explorers were so fixated on the acquisition of gold and land that they failed to see this "bewitching vegetable's" worldwide market potential.

Spanish explorers and their slaves, who soon set up shop in Central America, Brazil, and the West Indies, quickly adopted the local habit of a daily smoke. Available materials and customs of this area developed the prototype of our modern-day cigar, and evidence suggests that today's most commonly harvested types of tobacco originated in this area as well. Smokers would take the finely cut tobacco and roll it into a larger vegetable leaf such as that of maize. Further north in Mexico, Western visitors were learning the art of the pipe—often an ornately decorated tube carved from tortoise shell, silver, wood, or reed. Those who ventured up to and throughout North America, like Samual de Champlain, saw the pipes of the North American Indians which were of various shapes—as opposed to the straight tube type—with carved-out bowls for holding and burning the tobacco.

Above: Winston Churchill's love of a good smoke was so renowned that he had a cigar named after him.

> **THE ORIGIN OF TOBACCO**
>
> The Prophet Mohamet was walking in the country when he saw a serpent, stiff with cold, lying on the ground. He compassionately took it up and warmed it in his bosom. When the serpent had recovered, it said:
>
> "Divine Prophet, listen. I am now going to bite thee."
>
> "Why, pray?" inquired Mohamet.
>
> "Because thy race persecutes mine and tries to stamp it out."
>
> "But does not thy race, too, make perpetual war against mine?" was the Prophet's rejoinder. "How canst thou, besides, be so ungrateful, and so soon forget that I saved thy life?"
>
> "There is no such thing as gratitude upon this earth," replied the serpent, "and if I were now to spare thee, either thou or another of thy race would kill me. By Allah, I shall bite thee!"
>
> "If thou hast sworn by Allah, I will not cause thee to break thy vow," said the Prophet, holding his hand to the serpent's mouth. The serpent bit him, but Mohamet sucked the wound with his lips and spat the venom on the ground. On that very spot there sprung up a plant which combines within itself the venom of the serpent and the compassion of the Prophet—men call it tobacco.

Tobacco through the Ages

Back home in Europe, curiosity over this strange habit was aroused by Columbus' importing of the leaves in the late fifteenth century, and further piqued by exotic tales of foreign lives and the "naughtiness" of it all. Soon enough, those privy to this information had to have a smoking habit of their own. Thereafter, great ships brought seeds across the Atlantic and herbalists started to grow them in their gardens. Belgium had one of the earliest plantations in 1554. The smoking method of each European country—pipe or cigar—was largely dependent on the region they had taken it from.

Tobacco Becomes A Household Name

Jean Nicot, the French Ambassador to Portugal, took a keen interest in this new phenomenon and quickly sent off a letter to Catherine de Medici, the queen of France, hailing tobacco as the wonder drug of the future. Despite his predecessor Andre Thevet's earlier failed attempt to turn France onto this herb, Nicot's sell was victorious. He claimed it had medicinal benefits beyond all imagination, curing everything from melancholy to cancer. The royal family bought it, tobacco became Nicotiana (in Nicot's honor), and the herb went careening toward unprecedented success in France.

By 1570, physicians employed it in the treatment of virtually every ailment known. It was crushed into powders, drunk with teas, wrapped around wounds, stuffed in every orifice of the body, and revered by all. In truth, this "holy herb" was probably accountable for many fatalities, but it was not until years later that scientists and herbalists researched its properties and found it to be poisonous in certain situations.

In addition to its pharmaceutical use, smoking a tobacco pipe for pleasure was a budding fancy in the upper class-

By the late fifteenth century, Columbus was importing tobacco from the New World, and a new European habit was forming.

TOBACCO THROUGH THE AGES

Left: Portrait of Claude Monet, by Pierre-Auguste Renoir, 1872.

Above, top: Election Entertainment, by William Hogarth, detail.

Above, bottom: Nineteenth-century meerschaum pipes and tobacco tins.

es of continental Europe. However, from under the flurry of feet scurrying to the nearest source, rumbled a growing wave of condemnation. Originally, only the clergy expressed distaste with the "hedonism." Tobacco's opponents fell into two camps: the first, predominantly the clergy, felt it was a substance from the devil not to be used in any form; and the second felt that as a curative, the weed should not be exploited so lightheartedly.

In the meantime, European countries were broadening their sights on the globe and slipping into ports ever further from home. Italian, Portuguese, English, and Dutch seafarers sailed to distant lands with dreams of paradise in their skulls and a smoking roll of tobacco between their teeth. Curious about this oddity (just as Europeans were) and motivated by the social nature of trading, inhabitants of Asia and Africa adopted the habit. By 1605, smoking tobacco was an internationally recognized, although not yet commonly used, pastime.

Pipe Smoking And Plantations

Sir Walter Raleigh was living the good life in England in the early 1600s. He was in high favor of Queen Elizabeth, and was awaiting news of the tobacco-growing colony he had set up in Roanoke, Virginia (modern day North Carolina). Unfortunately, the colonists returned to England half starved and with stories of the low quality of the tobacco in that area. Raleigh's love of a pipe was unwavering, however, and his habit was emulated far and wide because of the position he held and the glamour

"I tried to stop smoking cigarettes by telling myself I just didn't want to smoke, but I didn't believe myself."

—BARBARA KELLY

The Curative Powers Of Early Tobacco

At various points in time, tobacco was prescribed by doctors to treat all forms of physical and mental ailment including asthma, dropsy, ringworm, scabs, scrofula, old sores, ulcers, all wounds, contusions, bruises, excessive phlegm, venomous, bites, breast afflictions of all kinds, venomous carbuncles, chilblains, flatulence, labor pains and the gestation period, halitosis, headache, helminthiasis, rheumatism, tumefactions, toothache, poisoning from venomous herbs, scurf, excessive bleeding, abscesses, wounds from poisoned arrows, colds, internal congestion stomachache, constipation, kidney stones, ozena, insect stings, cancer, colic, gout, indigestion, rabies, sciatica, spleen affections, surfeit, syncope, cataracts, dysentery, womb maladies, facial inflammations, itchiness, tumors of various kinds, old coughs, falling of the nails off the fingers, tonsillitis, convulsions, epilepsy, burns, deafness, dim eyesight, ileus, consumption, liver complaints, corns, warts, hemorrhoids, fevers, nasal hemorrhages, and gonorrhea.

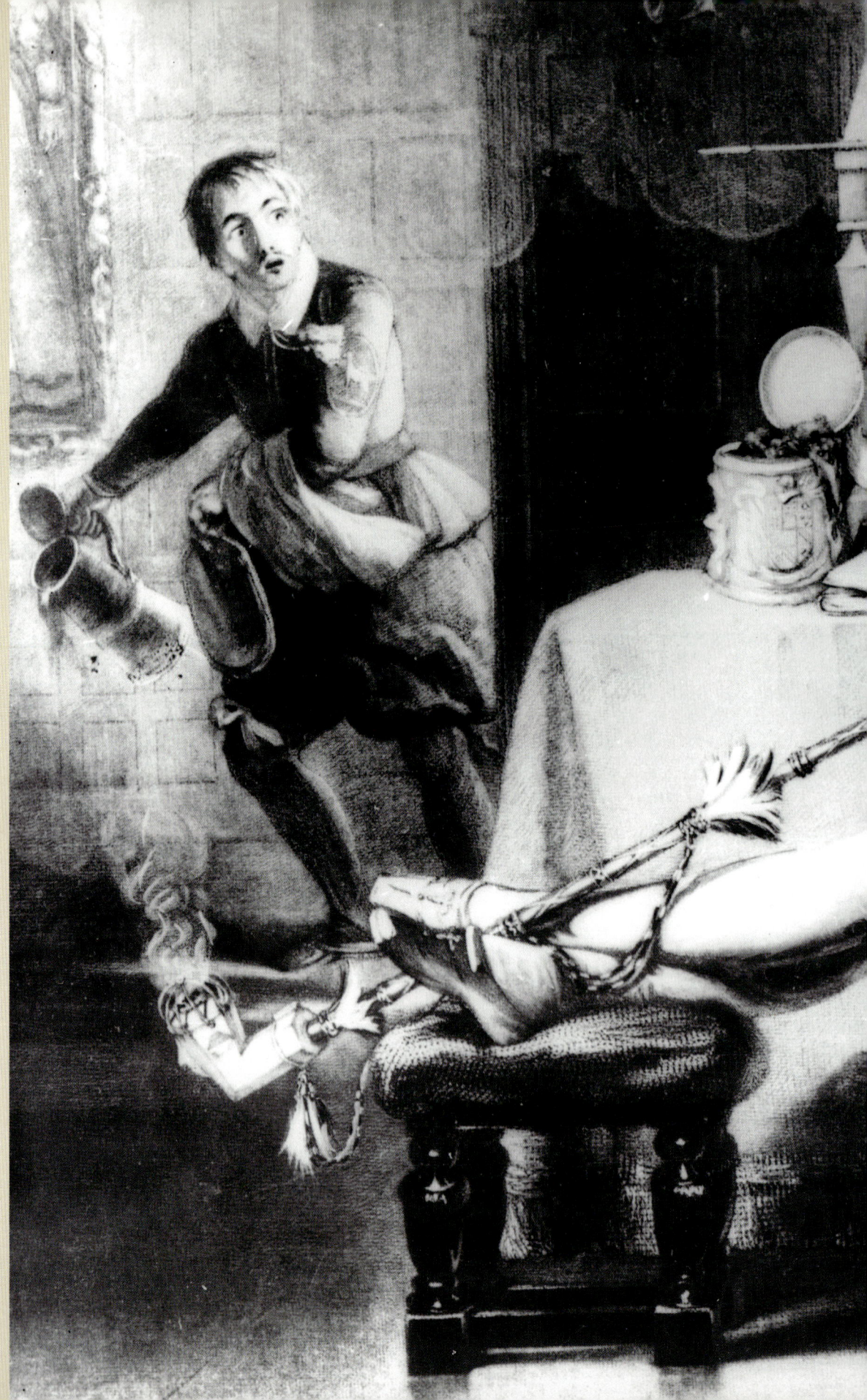

Sir Walter Raleigh, credited with introducing tobacco to England. Legend has it that when Raleigh first lit up at home, his servant, thinking him on fire, ran for a water jug and attempted to extinguish him.

Tobacco through the Ages

his pipe exuded. The pipe had finally swept England, and the rage alone was intoxicating.

John Rolfe, an enterprising colonist, seized this opportunity, imported tobacco seeds from Trinidad (then producing the best quality leaf) to Roanoke, and tried again. In 1610, the first non-native tobacco plants took root in the modern-day United States. Rolfe also—for economic rather than philanthropic reasons—established amiable relations with the natives of the area by marrying Pocahontas (whom the colonists had kidnapped), and secured a food source for the incompetent Englishmen while their new crop bloomed. From there on, tobacco became both the savior and the future of the English colonies.

Throughout the first decades of the seventeenth century, pipe smoking continued to be *the* fashion statement among elite male partygoers and extravagant male socialites in Europe.

Left: The Briarwood Pipe, by Winslow Homer.

Right: He may have been the master of disguise, but Sherlock Holmes' pipe was a dead giveaway.

Tobacco through the Ages

Expensive and exclusive paraphernalia was developed and entered into the unspoken competition for the most suave smokers. Except for the Spanish and French—who preferred snuff and cigars respectively—and the Swiss who remained wholly uninterested for a few more years, pipe-smoking reached an all-time high. The young American colony of Virginia reaped the benefits of this fad, too. By 1622, the farmers had been contracted as the exclusive suppliers of tobacco to a huge English and Irish market for a seven-year period.

Voices Of Opposition

Women did not take up the pipe during this boom, but it was only a matter of years and enough secondhand smoke before they asserted their rights to indulge. At the time, though, many joined doctors and priests in a vehement protest of the habit. Elsewhere in the world, rulers cracked down on this "Western evil" and forbid citizens to use it through varying degrees of threat. In Turkey, a tobacco smoker received the stem of a pipe up through the cartilage of his nose and into his brain! Chinese rulers threatened decapitation, and Russians who indulged were shipped to Siberia.

King James I of England was staunchly opposed to this trend and cited it as a filthy habit derived from people he considered to be "filthy and barbaric." In his self-published *Counterblast to Tobacco*, he describes smoking as "a custom loathsome to the eye, hateful to the nose, harmful to the brain, dangerous to the lungs, and in the black stinking fume thereof nearest resembling the horrible Stygian smoke of the pit that is bottomless." His disdain did little to temper the fervor, and before long his resolve collapsed in the face of a potential economic killing. Eventually, he gave moderate support

Left: "She's going to smoke!" When Edward VII, an avid cigar smoker, took the throne after the death of his tobacco-hating mother, he brought an end to the Victorian Era, declaring "Gentlemen, you may smoke!" The ladies, alas, were not included in this liberation, and were forced to continue smoking on the sly.

A Long Way Indeed

While many places of business today—including bars, concert halls, and restaurants—have banned smoking on their premises, in the past, smokers were welcomed at most places of work and play. (That is, of course, as long as they were men.)

In the 1920s, there were frequent cases in the United States of women who

were fired from jobs, expelled from colleges and universities, or otherwise punished for daring to smoke in public. The most famous case concerned a well-known opera singer, Madam Shumann-Heink, who is believed to be the first women to offer a public testimonial for a cigarette. Shortly after doing so, however, a number of her impending concerts were unceremoniously canceled. After learning this costly lesson, she denounced the use of tobacco altogether.

Tobacco through the Ages

THE NEED TO SNEEZE

There is a considerable art to smoking, a calming ritual that in and of itself soothes the smoker's psyche as much as it satisfies their craving. Yet snuff offers no such ritual: there is no artful cutting, as with cigars; there is no packing of the pipe. There is only the inhalation of a foreign substance, delivering an immediate, one-two punch. While avid snuff users believed it to be a cathartic which cleared the head, brightened the eyes, and invigorated the brain, the truth is that it is the sneeze (and the feeling of well-being that followed it), combined with the slight narcotic effect of the snuff, that made snuff a pleasure.

The taking of snuff also differs from smoking in a simple, but crucial, way: While smokers will slowly build up a resistance to certain aspects of the smoke, such as coughing, the snuff sneeze is invariable. Even the most avid snuffers failed to build up a resistance; no matter how often a pinch of snuff was inhaled through the nostrils, the snuffer was assured a fine, "cathartic" sneeze.

Sterling silver snuff box, c.1820s

to the American colonies in their agricultural pursuit.

A New Habit For The Elite

Towards the close of the 1600s, as the pipe's popularity exploded among the middle and working classes, the elite once again went in search of a habit to call their own. Snuff had long been favored in the French courts, and was quickly and wholeheartedly adopted by the upper classes throughout the rest of Europe. The 1700s earned snuff a place in tobacco's history and relegated the pipe to the dark corners of "low-class" establishments.

High society demanded extravagance, and so elaborate snuff boxes and associated materials emerged. In order to snuff properly, one must have owned not only an ornately decorated and minutely detailed box, but a wooden grater with a trough at one end to catch the snuff, a pin to clear the holes of the grater, a rake for separating the snuff, a spoon for taking the snuff, and a rabbit's foot to dust the upper lip. Some people even carried their snuff in the head of their cane. During the reign of snuff, ladies finally demanded their due in the tobacco world. Their bold displays in public shocked traditional society.

The procurement of snuff became an artform, and many of the eighteenth-century methods are still employed today. Merchants developed snuff to satisfy even the most persnickety of noses. The leaf preparation, fermentation and drying, grinding (called mulling), sieving, adding of essences, and maturation could take up to six months. It became available in a variety of granule sizes, flavors, and scents.

Masulipatam, scented with Oriental spices, is still a popular variety among modern-day snuffers.

The Resurgence Of Smoke

Snuff's heyday came to an end in the 1830s. Victorian England had much less tolerance for the stained handkerchiefs and drippy noses of the earlier years. Once again, it was back to smoking. This time, it was in a form that the Spanish had seen in their early exploration of the Americas and had never given up—cigars and soon afterwards, cigarettes.

The cigar was one of the first seen methods of tobacco use. The Spanish readily adopted it from their colonists in Central America in the 1600s. Perhaps if the pipe had not taken off with such a surge, cigar smoking would have seized Northern Europeans earlier. As it was, by the time Great Britain and Western Europe came around to cigar smoking, most of the

Cigarettes and soldiers have gone together since the American Revolution. During World War I, soldiers in the trenches found it safest to light only one or two cigarettes at a time, so as not to give night snipers a well-lit target at which to aim their bullets. The practice gave birth to the superstitious belief that lighting "three on a match" was a sure bet for bad luck.

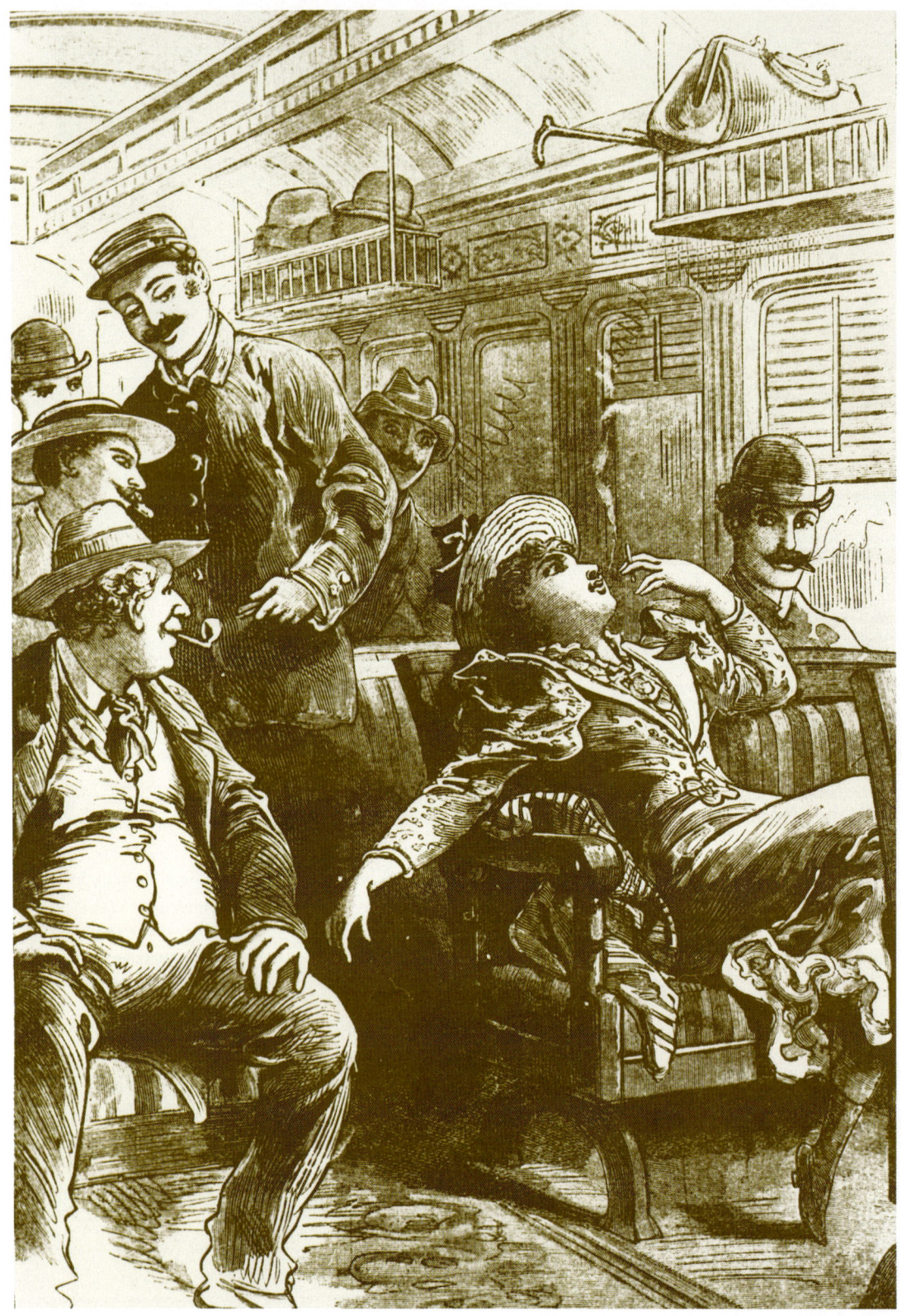

world was well ahead of them. The Spanish had introduced the cigar to Asia when they were in the Philippines, and it had been spreading steadily there and throughout Russia and Turkey. When smoking was again sanctioned in London clubs, the stage was set for the next social explosion: the cigarette.

During the Crimean War (1854-1856), French, English, and Russian troops fought alongside one another and shared smoke with the Spanish. The Spanish, for their part, had been working for decades on the perfect "pepelete" in which to roll their tobacco, and introduced it, in turn, to their compatriots. The soldiers brought these tiny sticks of tobacco, rolled in thin paper, back to their respective countries to launch a smoking phenomenon and social debate unprecedented in history and still alive today.

A Fashion Statement And Economic Gold Mine

Never were the laws of supply and demand so clearly displayed as in this

Tobacco through the Ages

glowing new market. The cigarette fulfilled all the desires of the late 1800s. European urbanites in the throes of an industrial revolution wanted a quick fix and had no patience for the frivolities of the past. They found the vogue irresistible. The pipe and snuff quickly became impractical and undesirable, and the cigar—while still appreciable—could not be forced into a speedy day, so it was saved for the evening.

There was no class or age division with the cigarette; everyone indulged heartily and with increasing frequency. While for the most part society was still uncomfortable with female smokers, some lit up to further crumble the already flailing Victorian conservatism. Opponents came out in droves to launch a fiery protest against this evil habit that was sweeping the world. Entrepreneurs pushed up their sleeves and embarked upon creating one of the world's largest industries.

The first cigarette smokers rolled their own in the available thin papers, but this soon proved time consuming and unsatisfactory. On both sides of the Atlantic, businessmen realized the manufacturing dream of a lifetime, and started to produce machine-rolled cigarettes from Turkish (then very popular) tobacco. Once the Virginian tobacco industries were steadily producing enough to supply the market, that became the desired flavor. The United States quickly surged to the forefront of the market in both production and export, and asserted its independence by adopting chewing tobacco as a national identifier.

Tobacco Use
In The United States Of America

While Europe had seen many tobacco habits come and go from as early as the sixteenth century, American settlers had yet to spawn their own unique tobacco habit. With newly gained independence from Great Britain and a refueled disdain for the snobbery of their former continent, they resurrected an old habit of the land. They took their leaves between their teeth just as Amerigo Vespucci had seen

"Smoking is one of the leading causes of statistics."

—Fletcher Knebel

TOBACCO THROUGH THE AGES

Opposite: *You Talkin' to Me?*, by Kim McCarty, 1991.

when he reached Margarita Island, off the coast of Venezuela. Vespucci, for whom America was named, once described tobacco as a "green herb which they (the natives) chewed like cattle to such and extent that they could scarcely talk."

During the late 1800s, Europeans were using Virginian tobacco in their cigarettes and for their urban lifestyles. Americans, on the other hand, were mass producing chewing tobacco to accommodate their rough, on the move, land-navigating ways. According to *Tobacco and Americans*, by Robert K. Heimann, "of the 348 tobacco factories listed by the 1860 Census for Virginia and North Carolina, only seven were smoking tobacco producers, and only six of the quid establishments mentioned smoking tobacco as a sideline."

States west of the Appalachian Mountains, and specifically in the Ohio Valley region, had been cutting into Virginia's tobacco market. With their cultivation of Burley leaf and their subsequent production of White Burley, both of which can absorb large amounts of sweetening for flavor, a new distribution center was secured. In having licorice, seasoning, and even fruit extracts added, chewing satisfied even the sweetest tooth.

But like many of these tobacco trends, chewing's popularity rose and fell again within the fence of a single century. By the time the Industrial Revolution came to North American shores, American men no longer wanted to be free-spitting out on the range and American women were gearing up for their role in tobacco culture. A few dedicated smokers, like Mark Twain, paved the way for the cigarette's upcoming sweep of America.

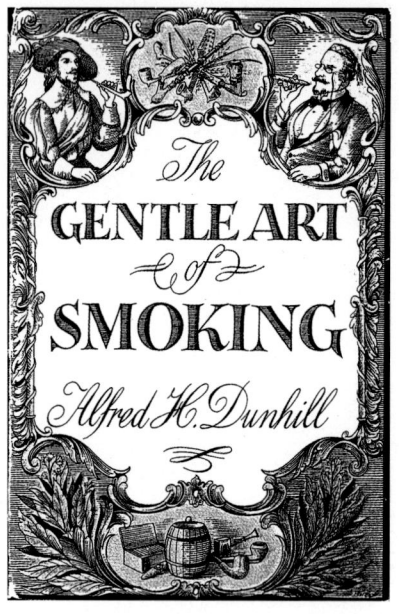

"He who lives without tobacco is not worthy to live."

—Molìere, *Dom Juan*

Right: The Pipe Smoker, by Paul Cézanne, c. 1895–1900.

CHAPTER TWO

THE PIPE

A World Of Different Pipes

Before the European socialites of the 1700s deemed the pipe supreme, their smoking predecessors had long been making tobacco pipes by hand. In his book, The Gentle Art of Smoking, *Alfred H. Dunhill provides a thorough history of the pipe from which the following information is taken. Alfred H. Dunhill's father founded Dunhill Pipes Ltd. in 1907, and it remains the maker of some of the world's most highly sought-after pipes even to this day.*

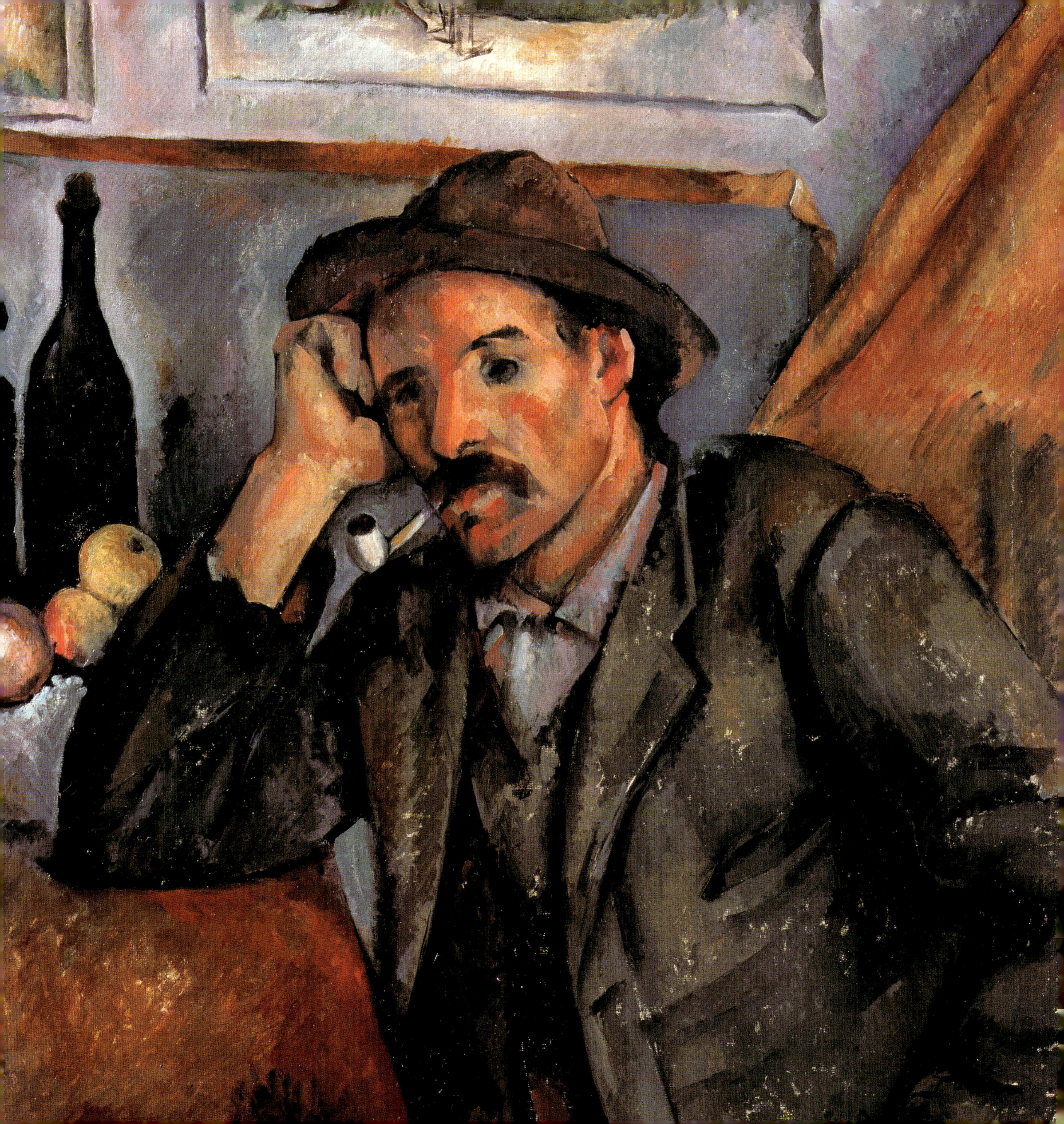

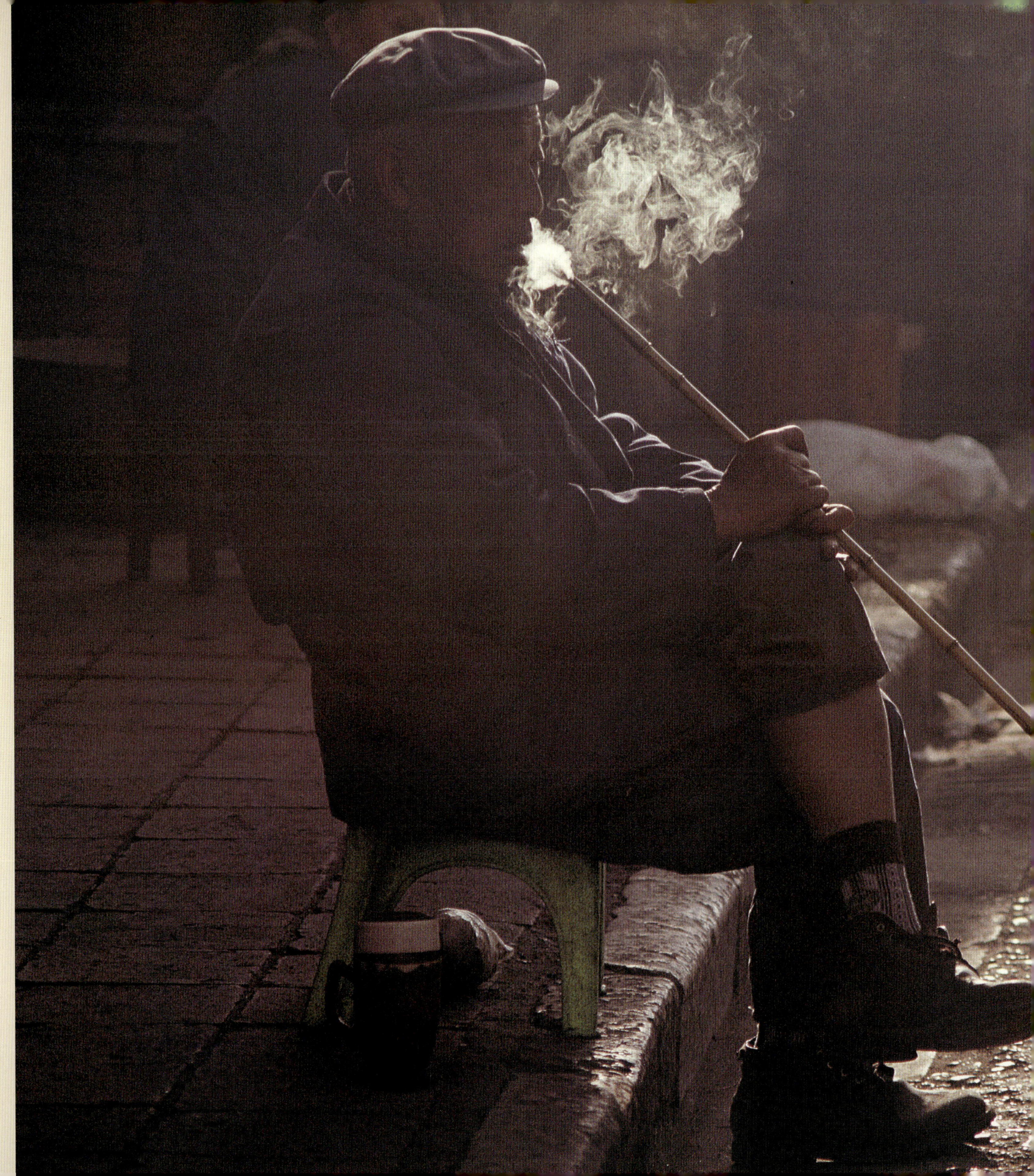

PROOF OF A LONG HISTORY

According to his report "Early Smoking Pipes in Africa, Europe, and America," archeologist Thurstan Shaw unearthed a number of clay smoking pipes during the course of an excavation in Ghana, West Africa, in 1961. Shaw estimated the pipes to be from between the end of the sixteenth century and the early seventeenth century. He found ninety-six pieces and fragments, and categorized them into eight types according to their shape, size, thickness, convexity, and external decoration. After considerable research into these pipes' antecedents, Shaw concluded that the Africans may have modeled the pipes brought home by England's Sir John Hawkins during his involvement in the slave trade in that area. If this is the case, it supports the belief that stem-socket pipe smoking was practiced in West Africa long before smoking became common in England. This comes as no surprise: Records of smoking in the Philippines pre-dates tobacco's introduction into Europe itself by more than thirty years. Once introduced on the Guinea coast, smoking spread rapidly into other parts of Africa, long before Europeans picked up the habit.

Left: **Man with opium pipe, Xian, China.**

SMOKE

THE PIPE

For thousands of years, the native peoples of West and Central Asia, as well as those from some areas of Africa, had used "earth-pipes." Earth-pipes consisted of a hole in the ground in which to burn the weed (usually hemp in these non-American cultures), and an underground "tube" created by drilling a stick down and toward the "bowl." This method was employed as recently as World War I by Native American soldiers.

Elsewhere, and more recently, pipes were carved out of any available materials and often included painstakingly minute detail. Leaf stems, reed, bamboo, crab legs, and even human bone are all part and parcel of the pipe's illustrious history. Many pipes from the Algonquins of North America have sacred animals carved upon them. Their shapes—in this case long and tubular—indicate the important role that pipes played in the religious and medicinal practices of the culture.

The calumet, or peace pipe (as previously mentioned), was one of the first aspects of American culture that European explorers came to understand. Oftentimes, the presentation of that pipe meant their survival in a land whose laws they had broken. These pipes were carved by hand and further decorated with horsehair or feathers.

After Portuguese and Spanish explorers brought tobacco to Asia in both pipe

Pipe smoking has a history all its own. *Opposite:* Chinese opium den, undated photograph. *Left:* Rajah of Chambra, Bahari, c. 1730. *Below:* Old clay tobacco pipes, Scotland.

The Pipe

Right: **Kramer's love of a fine Cuban cigar has caused plenty of trouble on T.V.'s *Seinfeld*, but actor Michael Richards prefers a pipe.**

Below: **This New Guinea native sucks on a traditional bamboo pipe, filling it with smoke. He'll then pass it on to his companions so that everyone can have a puff.**

and cigar forms, the natives of those lands adapted the smoking tools to fit their own culture. The people in and around China took to pipe smoking and decorated their bamboo or metal wares with ivory, jade, and lacquer. They also tended to make smaller bowls on the pipe to accommodate the opium they added to the tobacco. Many of the other Asian societies who had adopted the pipe-smoking habit, made their pipes out of clay. These, too, were intricately ornamented.

The unique pipe tradition developed in Borneo, spread throughout New Guinea, and eventually into the various clans of the Australian Aboriginal peoples. "The practice among the Papuans is for one person to fill the large stem with smoke and then, having removed the bowl and closed the open end of the tube with his hand, to pass the pipe to each member of the company who inhales the smoke through his nostrils. When the stem is empty of smoke, the bowl is replaced and the ritual begins again."

The water pipe, or dakka, had been widely used throughout Western Asia and parts of Africa, and comes in many forms. The basic structure, though, is a vessel of water from which extends a tube and bowl filled with tobacco. The smoke passes through the water for cooling and cleansing before moving up into the other tube from which the person is drawing the smoke. India, in particular, has greatly developed, improved, and made artistry of this type of pipe.

The many pipes from Africa are too diverse to be given justice here. The pipes were made of materials ranging from clay to wood and metal, and leather to various gourds and nuts according to area they were taken from. They are all shapes and sizes and

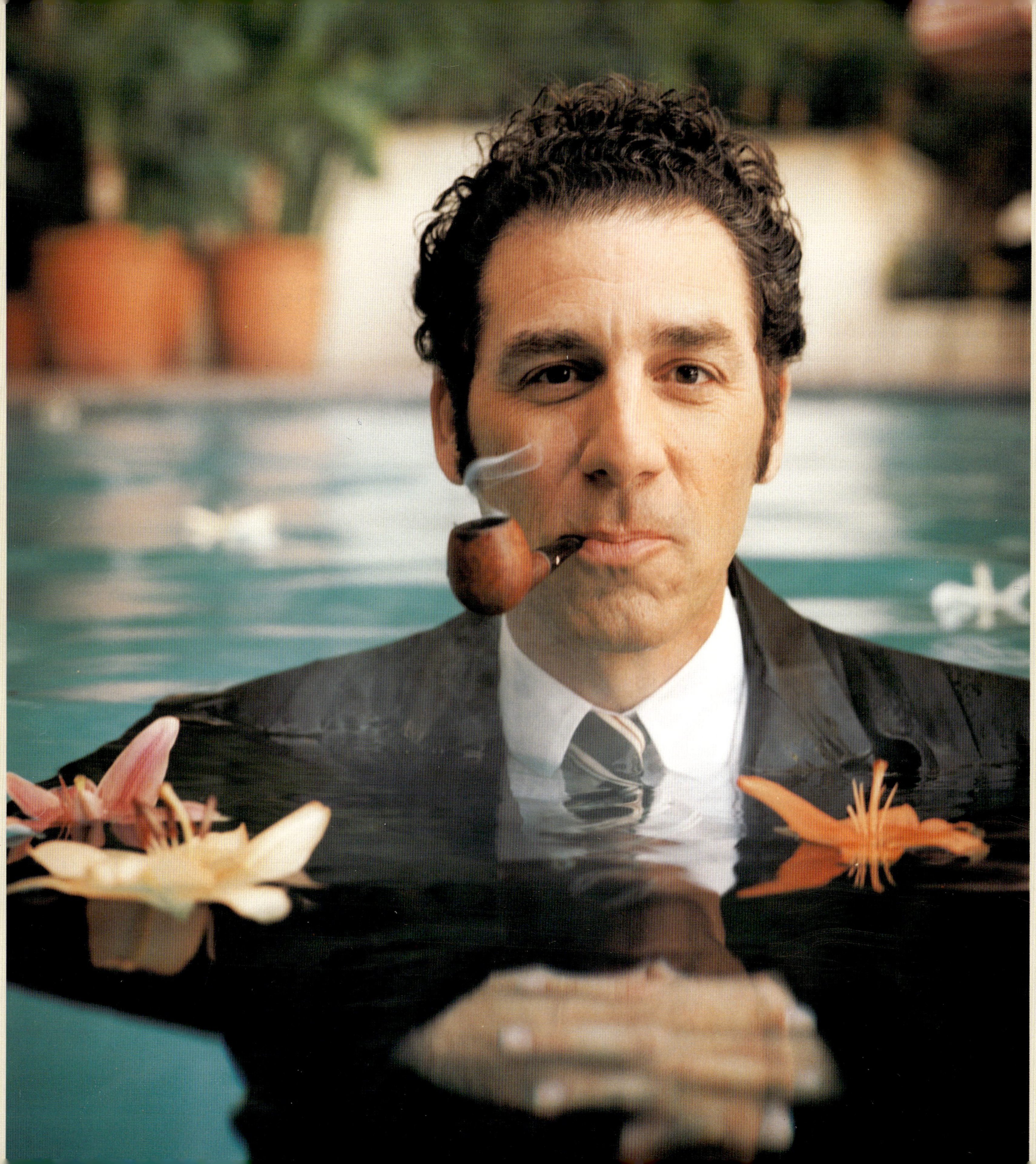

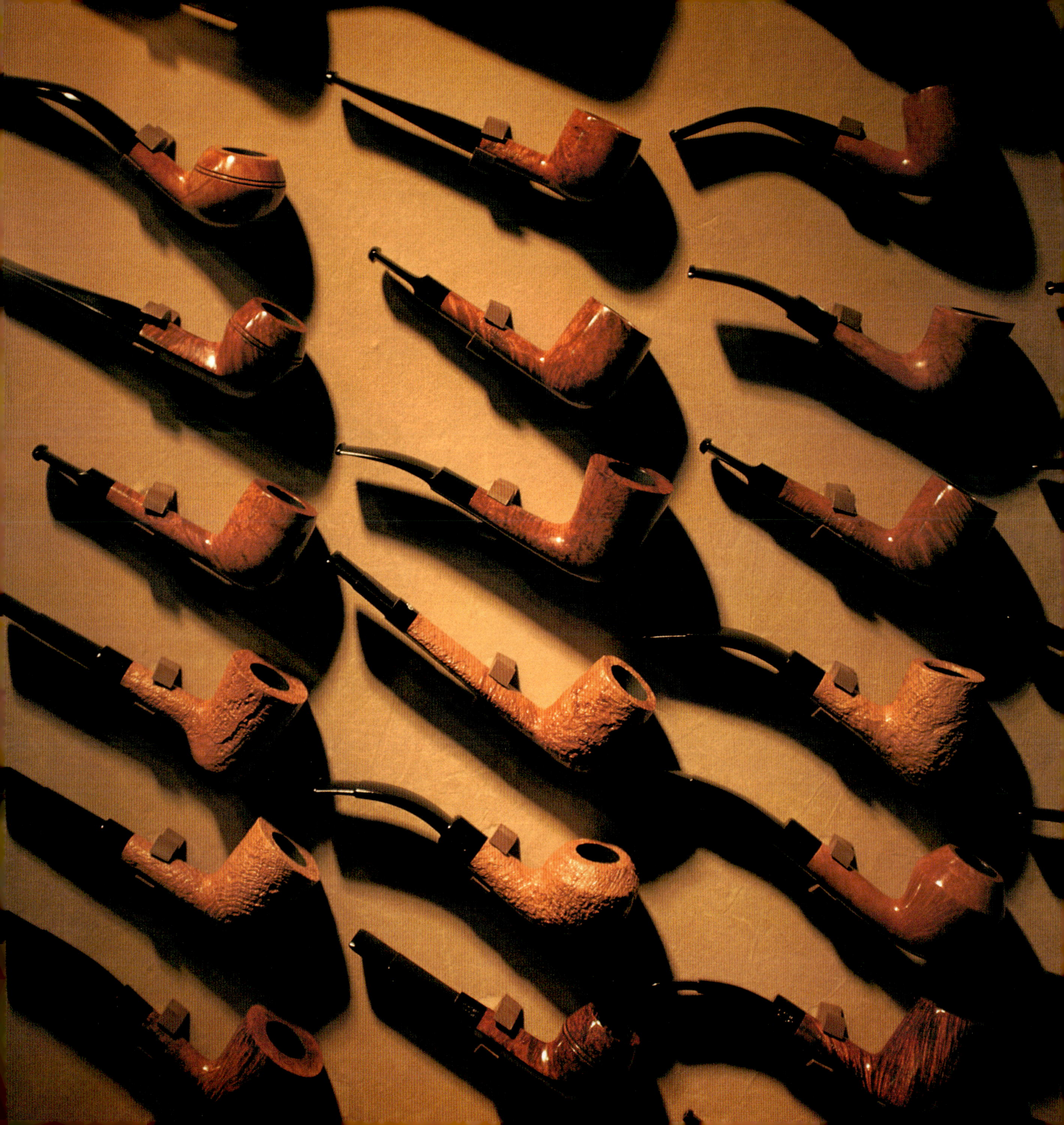

THE PIPE

are decorated with innumerable patterns, carvings, inlay, and colorings. While originally the pipes were used for hemp, the introduction of tobacco changed the continent, and the pipe mutated into different forms to serve various smoking tastes.

The Finest Of England And France

Had Sir Walter Raleigh been around long enough to see the trend he started explode as it did, he may have been very well pleased by the quality and diversity of the pipes his homeland produced. Once people saw the market in pipemaking, the industry began to thrive. Clay and porcelain pipes were being produced, sold, and used by the grossload. Given an Englishman's love of a pipe and clay's short lifespan, he went through his purchases very quickly. The designs were becoming more interesting and the demand for a durable and aesthetically pleasing pipe increasingly louder. Finally in France, the clay version was copied in the wood of a local species of heather and a beau-

Elementary, my dear Watson... Basil Rathbone breathed life into the celebrated sleuth Sherlock Holmes, a master of deductive reasoning and constant pipe smoker.

THE PIPE

SIR WALTER RALEIGH AND FRIENDS

The Mermaid Pub of London is well known as the gathering place for seventeenth-century London's great literary minds. According to history on one such meeting, Sir Walter Raleigh offered his friends—Ben Johnson, Beaumont, Selden, Fletcher, and Shakespeare—pipes and tobacco. Ben Johnson claimed, "Tobacco, I do assert, without fear of contradiction from the Avon skylark, is the most soothing, sovereign, and precious weed that ever our dear old Mother Earth tendered to the use of man! Let him who would contradict that most mild, but sincere and enthusiastic assertion, look to his undertaker. Sir Walter, your health."

Above: The Treason of Images, by René Magritte, 1948.

tiful and durable pipe—the Briar Pipe (from the French 'bruyère' for that wood)—met the demands of the world.

The making of the finest briar pipes may include up to eighty different stages of production. The selection of an ideal piece of root wood from the Mediterranean countries where the shrub is most abundant, and the arduous processes of producing the perfect bowl, stem, and mouthpiece are carried out by highly skilled professionals. Both Saint-Claude, France—the home of the first briar pipe—and London are reputed for producing the highest quality pipes. These pipe makers turn out the three varieties of briar pipes—

On film, Edward G. Robinson was often typecast as a cigar-chomping gangster—but he really preferred the refinement of a pipe.

THE PIPE

46 SMOKE

THE PIPE

Lighting up in style. Left: You're Welcome! (Self-portrait of the photographer lighting his pipe) by Clarence Leino, 1953. Right: multicolored matches for the creative smoker.

smooth, sandblasted (displaying a rough, gripable surface), and carved—for many of the world's smokers. If a pipe is approved at its final stage, its purchaser will become the proud owner of an exceptional, unique, long-lasting, and pricey accouterment for his or her tobacco.

A Turkish Specialty

Second to the briar pipe is the meerschaum pipe. Meerschaum is a mineral (magnesium silicate) found deep in the land of a tiny village in Turkey. Miners risk their lives to bring up this "seafoam" in order to meet the market's needs. Here, too, are some of the finest artisans of these intricately carved pipes which change color from white to a rich brown as they absorb tobacco over years of mellow smoking. An evenly tanned pipe is an item to be envied by collectors and aficionados alike.

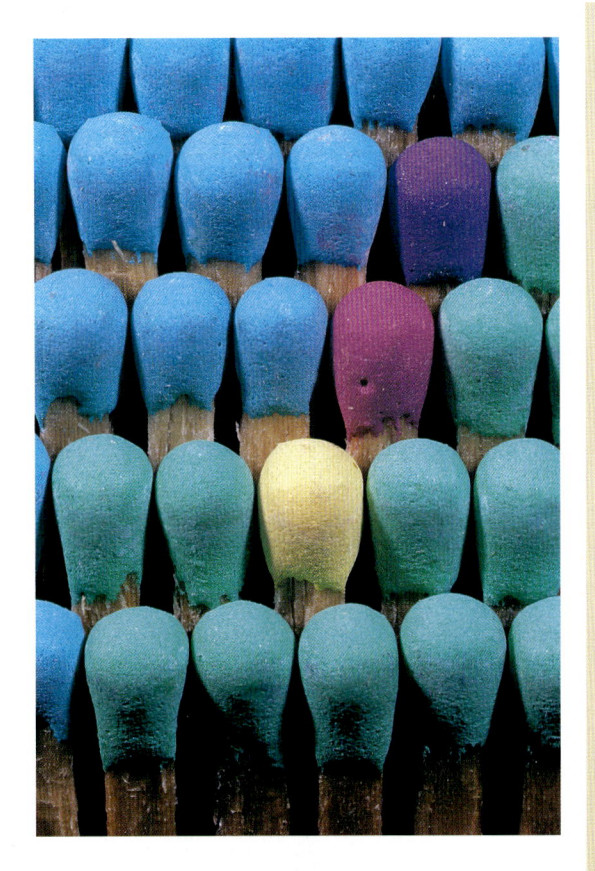

The Curing Of Tobacco

While there are dozens of different species of tobacco, it is Nicotiana Tabacum that fills the world's pipes, cigars, and cigarettes. Other species are added in moderation for flavor distinction or enhancement, but Nicotiana Tabacum from Central and South America has a naturally sweet taste and pleasant aroma. It makes up about three-quarters of the tobacco now smoked throughout the world.

MEERSCHAUM DEFINED

MEERSCHAUM, n. (Literally, seafoam, and by many erroneously supposed to be made of it.) A fine white clay, which for convenience in coloring it brown is made into tobacco pipes and smoked by the workmen engaged in that industry. The purpose of coloring it has not been disclosed by the manufacturers.

There was a youth (you've heard before,
This woeful tale, may be),
Who bought a meerschaum pipe and swore
That color it would he!
He shut himself from the world away,
Nor any soul he saw.
He smoke by night, he smoked by day,
As hard as he could draw.
His dog died moaning in the wrath
Of winds that blew aloof;
The weeds were in the gravel path,
The owl was on the roof.
"He's gone afar, he'll come no more,"
The neighbors sadly say.
And so they batter in the door
To take his goods away.
Dead, pipe in mouth, the youngster lay,
Nut-brown in face and limb.
"That pipe's a lovely white," they say,
"But it has colored him!"
The moral there's small need to sing —
'Tis plain as day to you:
Don't play your game on any thing
That is a gamester too.

—Ambrose Bierce,
THE DEVIL'S DICTIONARY, 1911

THE PIPE

"Pipe-smokers spend so much time cleaning, filling and fooling with their pipes, they don't have time to get into mischief."

—BILL VAUGHAN

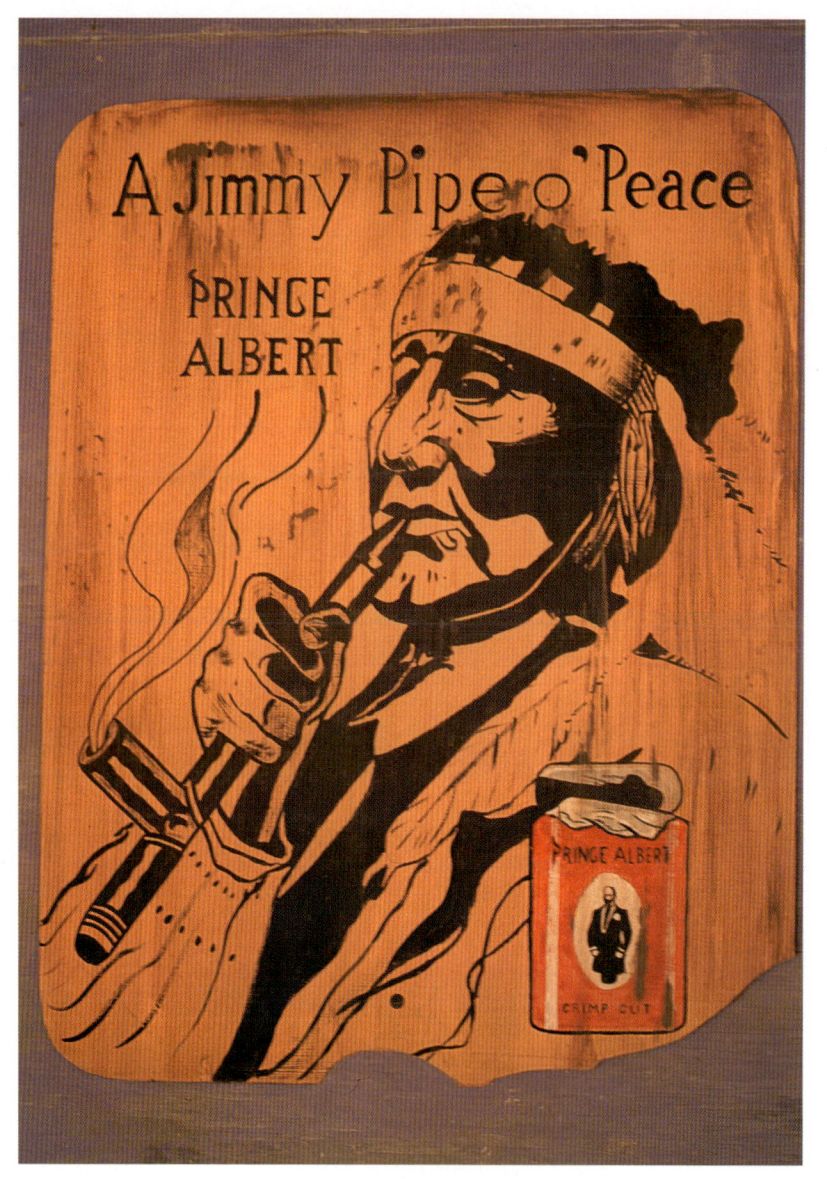

Left: Prince Albert pipe tobacco ad, c. 1875.

Right: Just another piece of the American dream: In the 1950s, it seemed like everybody's dad smoked a pipe.

As with any crop, tobacco plantations require fastidious maintenance. Individual farmers choose their planting and harvesting methods according to their area and climate, the leaf they are growing, and the market they are supplying. Once the plants are harvested, either by cutting the whole plant or by picking the leaves as they mature throughout the season, they must be cured or dried, and fermented. If the leaves are air-cured, they are hung in bunches and dry naturally. If they are flue-cured, the leaves are kept in a barn that is artificially heated to high temperatures. In

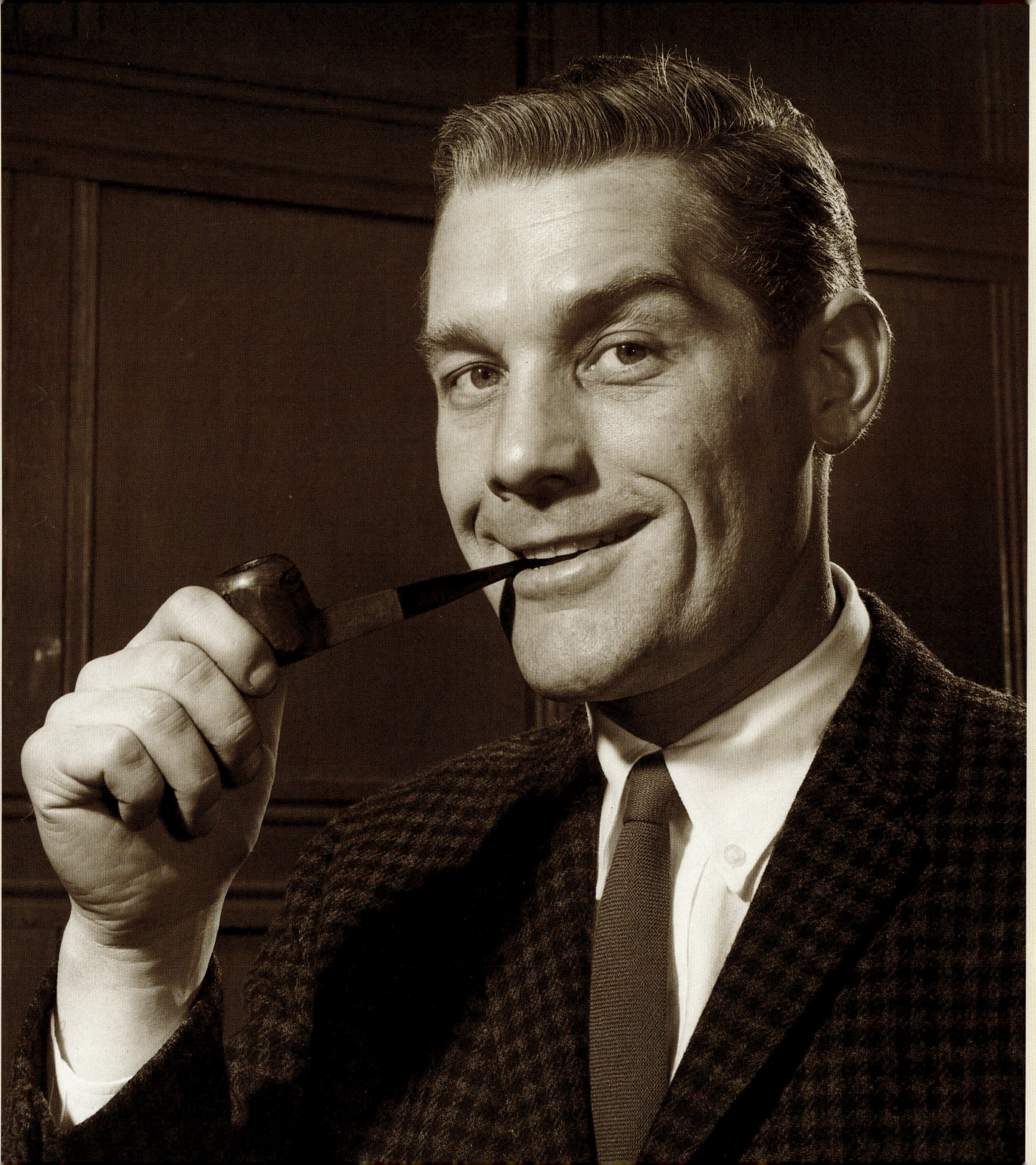

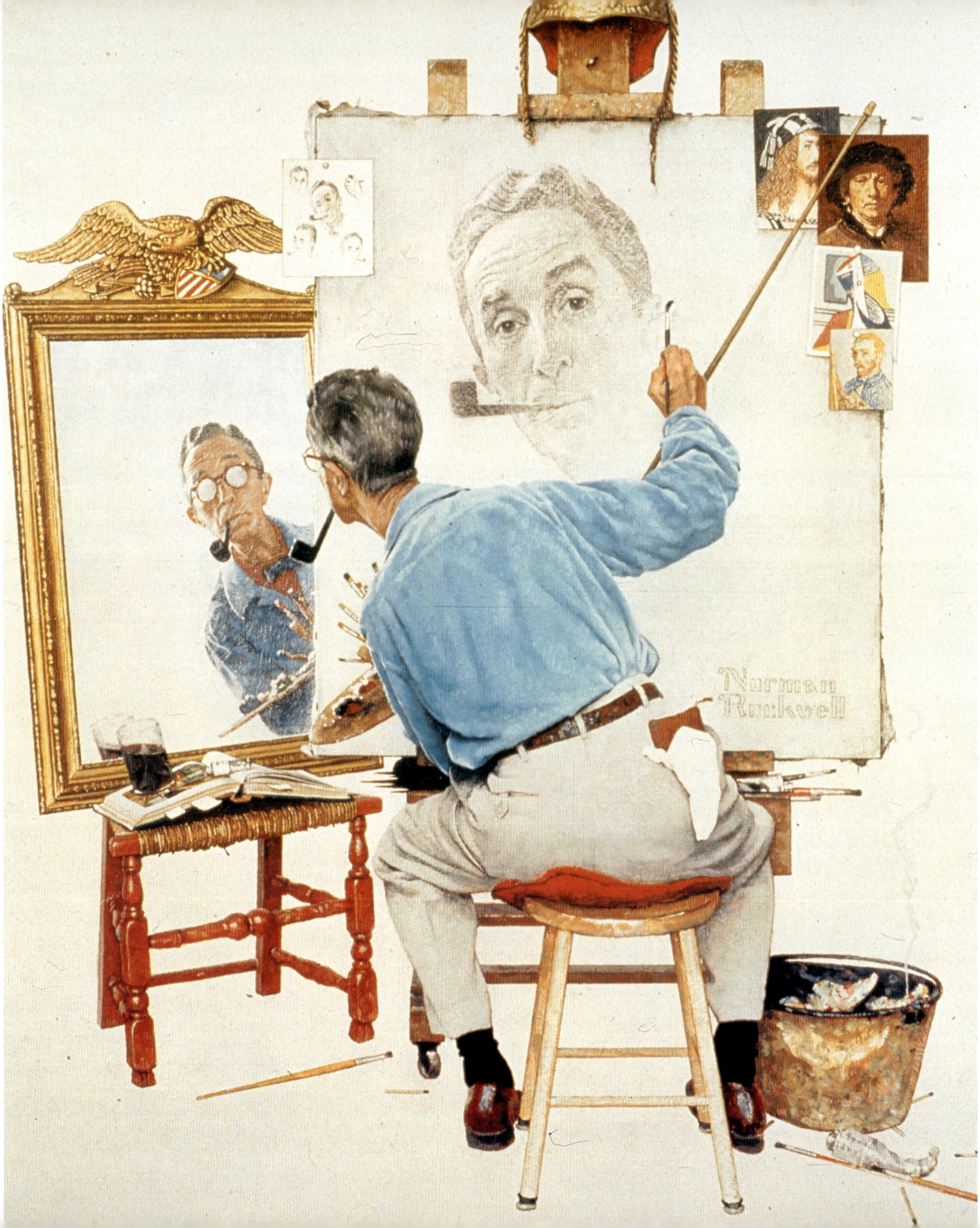
Norman
Rockwell

THE PIPE

the latter method, the farmer must keep a careful eye on the thermostat; temperatures too high, too low, or too inconsistent may ruin an entire harvest.

After the cured leaf is graded according to its quality and the part of the plant it's from, it needs to be aged properly. Tobacco is aged or fermented for up to three years or as little as six months depending again on the leaf and its projected use. To achieve a full flavor and quality aroma, the leaves are

Above: **Bales of tobacco aging: it's a long, slow process, but what a difference it makes.**

Left: Triple Self-Portrait, **by Norman Rockwell: three artists and three pipes.**

In Praise of the Pipe

Three hundred years ago
   or soe,
One worthy knight
   and gentlemanne
Did bring me here,
   to charm and chere,
to physical
   and mental manne.
God bless his soule
   who filled ye bowle,
And may our blessings
   find him!
That he not miss
   some share of blisse
Who left so much
   behind him.
         —Bernard Barker

Right: **The smart guy always prefers a pipe: Pierce Brosnan clamps down while explaining it all in** *Mars Attacks!,* **1996.**

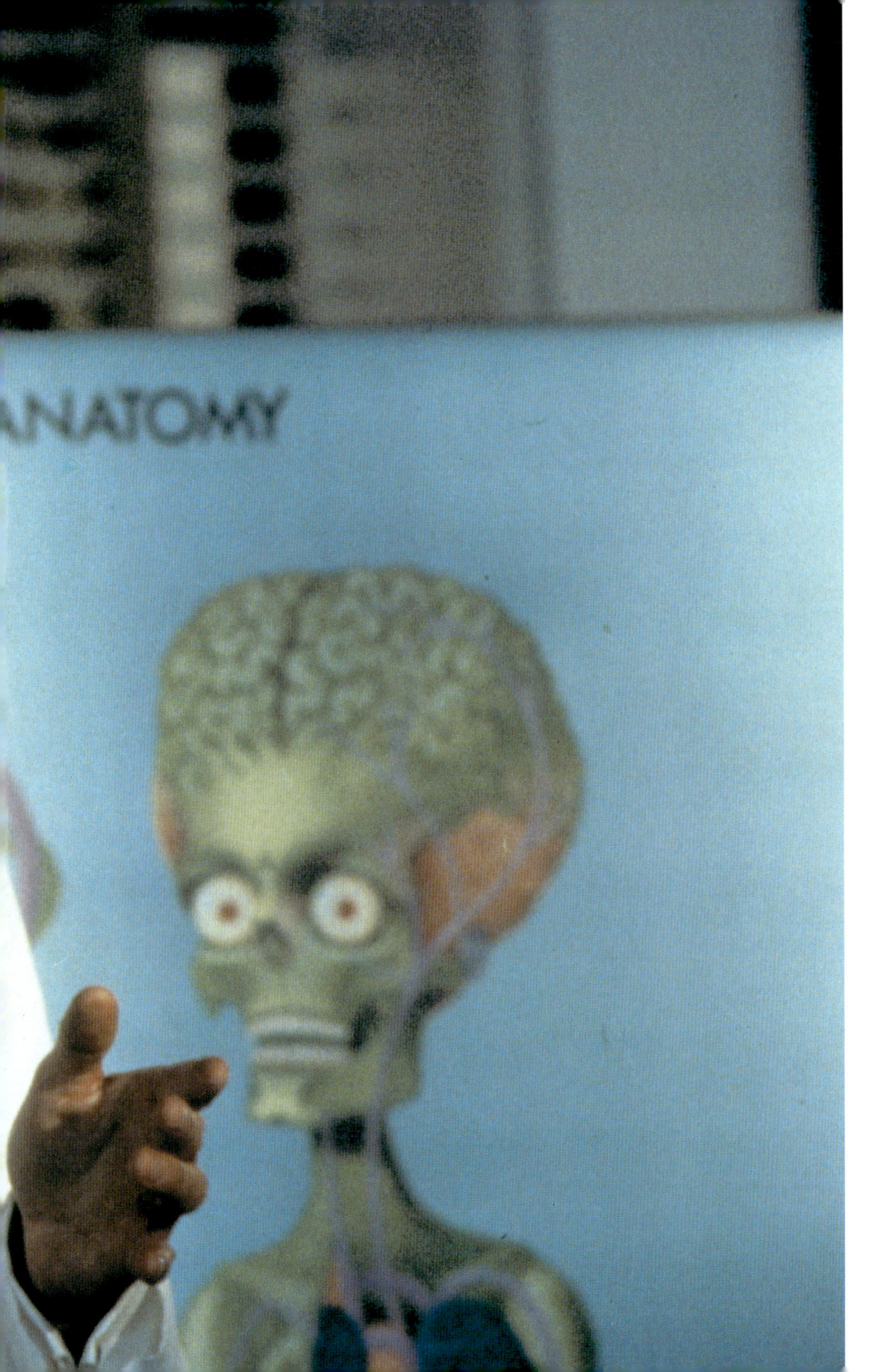

"The wretcheder one is, the more one smokes; and the more one smokes, the wretcheder one gets—a viscious circle!"

—George Louis Palmella
Busson du Maurier

The Pipe

Pipes—Not For Kiddies Anymore

The anti-smoking lobby has had a measurable impact on children's toys and entertainment. Some people think it's for better, while others aren't quite sure what to think. For a generation that made pop-culture heroes out of pipe-chomping Popeye, that didn't consider a snowman complete until it had the corn-cob pipe dictated by the song about Frosty, it may come as a surprise that their children's versions of toys they had grown up with are missing this important accessory. Modern versions of the classic children's toy Mister Potato Head are pipeless. In cartoons and movies, the good guys never smoke anymore. While many parents are happy to see this vice removed from their children's line of sight, some secretly wonder what's so fun about a non-smoking potato.

Top right: **The all-important accessory adds the finishing touch to a jack-o-lantern in this vintage greeting card.**

Bottom right: **Sherlock Holmes examines a handprint through the smoke, 1903.**

packed with varying degrees of moisture and remain thus until the manufacturer unpacks the fully fermented leaves at their proper time.

Individual manufacturers sometimes add flavors or sweeteners called "casings" to their tobacco before the final crucial step of "blending." It is this blending of different leaves that gives each brand of tobacco its stronger or milder taste, unique aroma, and burning rate. It would take a pipe lover years to smoke even a portion of all the tobaccos available. This, however, is part of the pleasure. Of the two types of pipe tobacco—namely Aromatic (flavored) and English (only pure tobaccos, up until recently, flavoring was illegal in Great Britain)—Americans prefer, overwhelmingly, the Aromatics.

To Be A "Real" Pipe Smoker

Pipe smokers use a number of accessories for their habit. A tamper is used to pack the tobacco down into the bowl of the pipe. These, like all smoking paraphernalia, come in a full range of styles and

54

Smoke

"The unfortunate thing about this world is that the good habits are much easier to give up than the bad ones."

—W. Somerset Maugham

British novelist W. Somerset Maugham indulges in a bad habit.

THE PIPE

"Sublime tobacco! which from east to west

Cheers the tar's labor or the Turkman's rest."

—LORD BYRON

prices. Pipe cleaners are a mandatory mate as well, absorbing the moisture from within the stem of the pipe. A tobacco humidor should be used for storage; it keeps the weed moist and fresh for smoking. And, of course, there is the mandatory pouch for travel from the easy chair.

The experienced smoker may want a pipe reamer to trim the buildup inside of the bowl down to an appropriate amount. All these purchases and preparations would be in vain if one does not have a reliable source of fire, so the most conscientious pipe smokers keep a large stash of reliable matches close at hand.

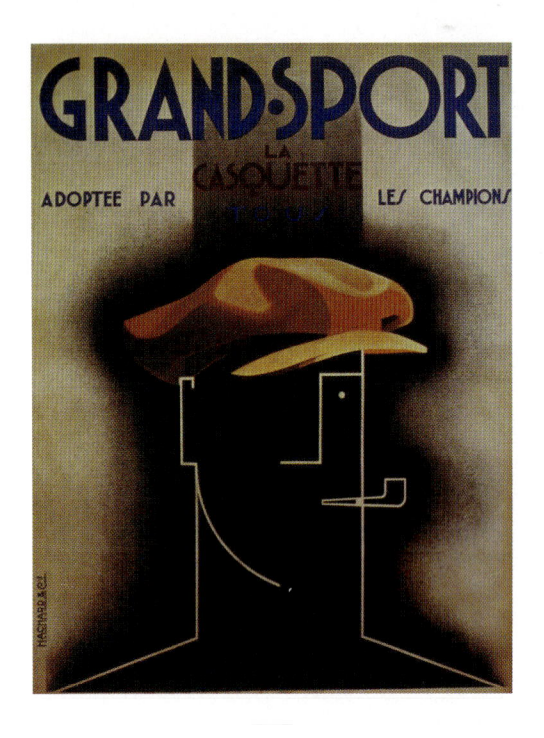

A Meal In Every Bowl

Much victuals serves for gluttony

To fatten men like swine;

But he's a frugal man indeed

That with a leaf can dine,

And needs no napkin for his hands,

His fingers' ends to wipe,

But keeps his kitchen in a box,

And roasts meat in a pipe.

—Samual Rowlands

Smoke

L.C.Smith & Bros.
JULY 1, 1908
REGISTER
OF
MONEY ORDER
POST OFFICES
LIST OF
SHOWING

Left and right: Once upon a time, cigars were a gentlemen's domain; nowadays, the ladies are in on it, too. Demi Moore (in one of her less shocking pictorials) graced the cover of *Cigar Aficionado* magazine, stogie in hand.

CHAPTER THREE

THE CIGAR

"If I cannot smoke cigars in heaven, I shall not go."

—MARK TWAIN

Times Shows Quality

In all the turbulence of smoking trends over three hundred years, no pleasure has enjoyed such longevity and steady growth as the cigar. From its creation at least one thousand years before the birth of Columbus, to its current haunt in popular culture, the cigar has represented precisely the attitude of outstretched legs, relaxed minds, and unburdened clocks that our rapid world often renders obsolete. While United States cigarette consumption has remained steady after years of decline, cigar use has increased enormously. Today, about three million Americans smoke more than four billion cigars per year.

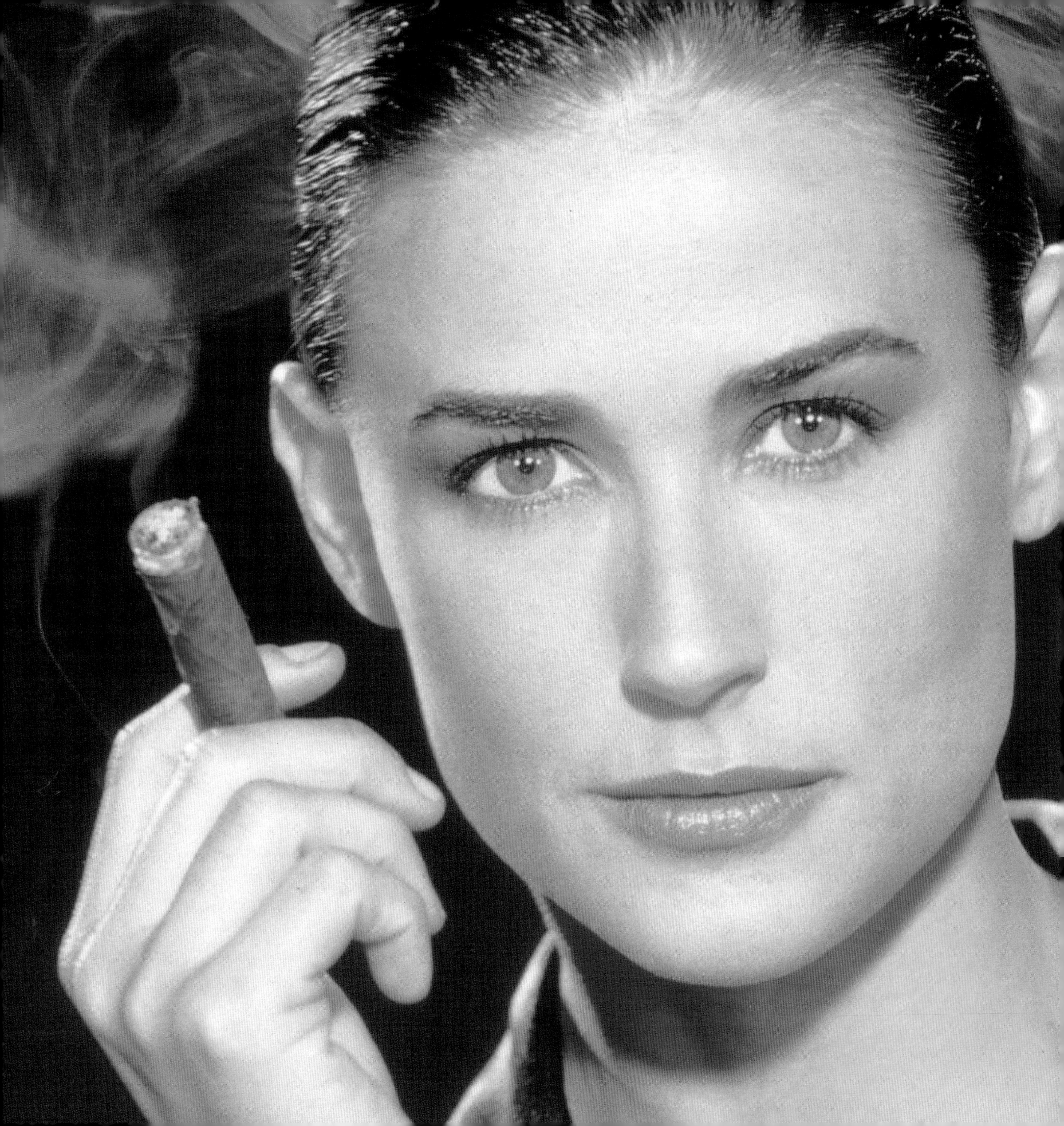

There are a lot of cigar-chomping comics, but there was only one Groucho Marx. With a wiggle of his eyebrows and a shake of his ash, Groucho delivered a punch line with punch, bringing his vaudevillian schtick to Hollywood and becoming a legend of the silver screen.

The Cigar

As mentioned, it was the cigar that first captured the sailors and merchant explorers in the "New World." Spaniards readily adopted the habit, but the upperclasses of other European nations wanted to separate themselves from the "lowlier" seafarers. Originally they chose the pipe and subsequently, snuff. While the cigar enjoyed a steady ebb and flow of popularity from its introduction, the late nineteenth century catapulted it to European society's—and to a temporarily lesser extent, American society's—social frontline.

All the fine homes of England and France soon converted one of their rooms into a smoking room, or divan. Like its Turkish namesake, this room was ornately decorated and embellished with the trendiest wares of the time. Men would congregate here in their velvet smoking caps and jackets to imbibe in smoke and gossip. The more daring women surely had hallways, conversations, and cigars of their own, but history prefers to keep that silent.

The Perfect Leaf

By that time, Spanish conquistadors had long been aware of the exemplary quality of the tobacco grown in one of the territories they had pillaged and they also knew that Cuba was the envy of the world. During the formative years of European cigar smoking, Spain took advantage of Cuba's exceptional

"A woman is only a woman, but a good cigar is a smoke."

—Rudyard Kipling

Above: Cigar Landscape, by Jacques Halbert, 1996.

THE GIFT UNWRAPPED

The cigar is a story in itself, a tidily rolled up package filled with layers of fine quality ingredients. But unlike a gift, when it comes to cigars, the wrapping turns out to be just as important—if not more so—than what's inside. And wrapping is itself a highly respected skill.

The inside of the cigar consists of the filler, a unique blend of flavorful leaves, blended and assembled to run the entire length of the cigar. This is the heart of the smoke, from which the flavor and aroma are derived. The filler is held together by the binder, a crude inside wrapper.

The final—and arguably most important—component is the wrapper. Leaves from locations as diverse as Cuba and Connecticut are carefully grown and harvested, and finally chosen for quality, size, and texture. The quality of the wrapper and the way in which it is rolled determine the appearance of the cigar, as well as the way in which it burns.

"Happiness is:

A good martini, a good meal, a good cigar and a good woman . . . or a bad woman depending on how much happiness you can stand."

—GEORGE BURNS

George Burns with the two great loves of his life: his wife Gracie Allen and an El Producto cigar.

THE CIGAR

Opposite: **Nicolas Cage, one of Hollywood's coolest leading men, lights up a stogie at the Cannes Film Festival in 1995.**

Right: **Women in Seville, Spain, roll cigars at the turn of the century.**

Below: **The world of high-fashion may never be the same: Supermodel Linda Evangelista sports a different kind of glamour in 1995.**

tobacco leaves by shipping them to Seville for their own production use. However, in the beginning of the nineteenth century when connoisseurs realized the quality of the tobacco was being sacrificed in the shipping, production began in Cuba. This move drastically affected both the cigar market and Cuba's own economic position.

Today, Cuba is still the center of cigar culture, producing the world's finest and most popular cigars. This is due to the unique soil conditions of the land—which greatly affect the final flavor of the leaves—combined with the unmatched expertise of Cuban torcedors, or cigar makers. Despite the worldwide panic among cigar smokers during Cuba's shift to communism in 1959, the government has supported the old methods of production thus preserving quality.

Throughout years of research and development, scientists have finally forfeited to the knowledge that the finest

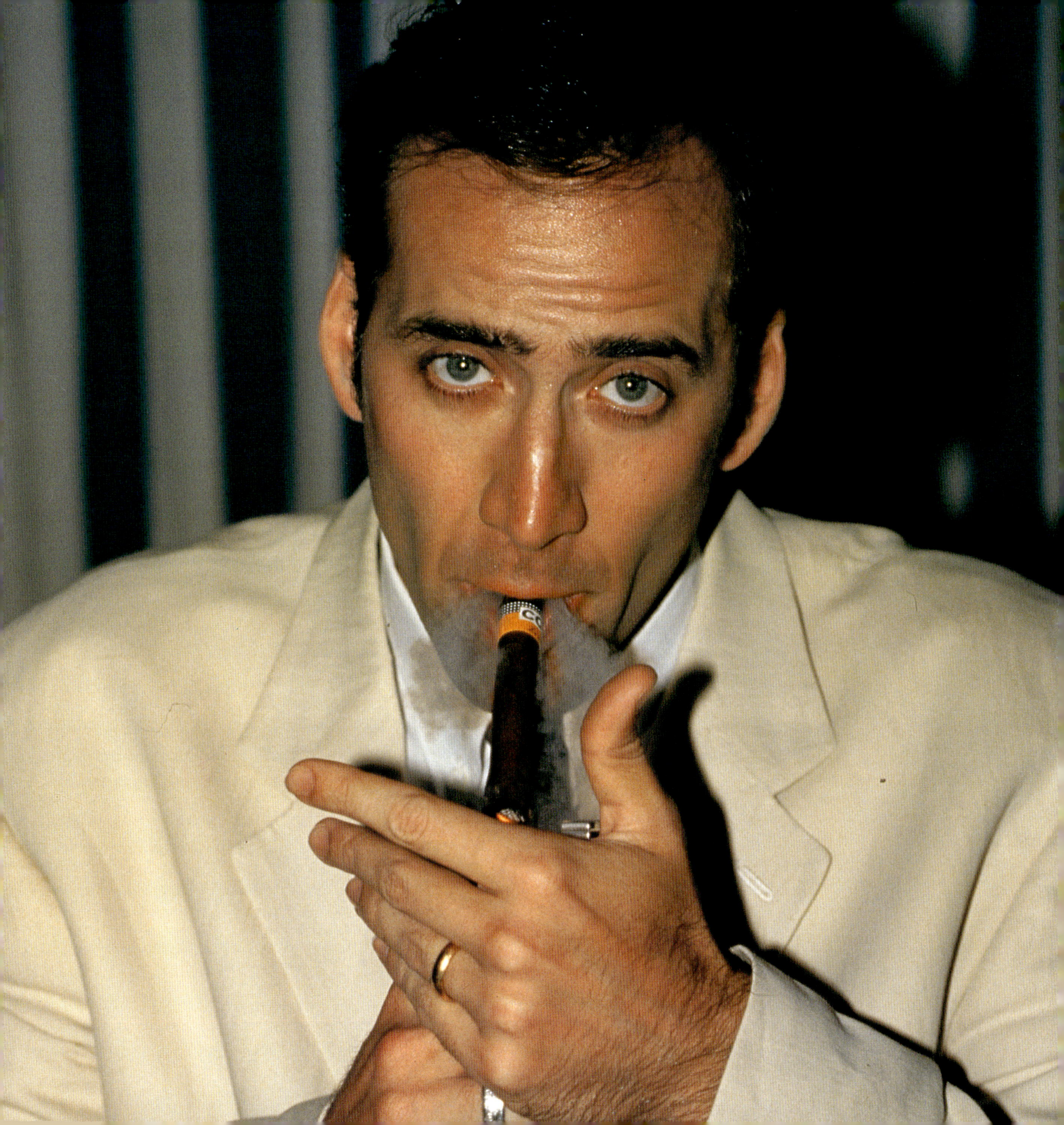

THE CIGAR

"What this country needs is a really good five-cent cigar."

—United States Vice President Thomas Riley Marshall, while presiding over the Senate

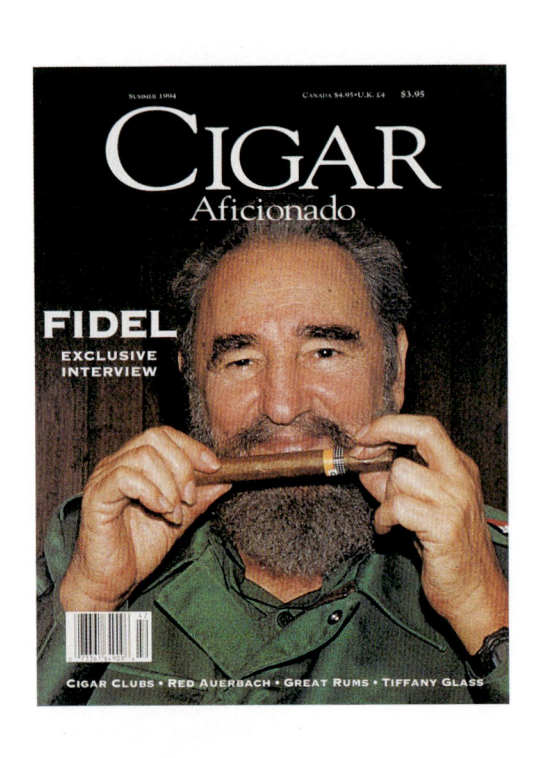

seeds, the most caring hands, and even hundreds of years of curing know-how, cannot affect the tobacco plant as much as the unique qualities of the soil in which it is planted. The red clay dirt of Cuba's Vuelta Abajo (Lower Valley) is truly the fertile crescent of tobacco farmland. Elsewhere in the country a fine leaf is produced, but it is here that the leaves of the puro—the messiah of cigars—are nurtured.

An Artisan's Work

Cuban torcedors do not learn their trade easily, and because of this, they are highly respected in their society. After a year-long apprenticeship, they must work for at least two years before earning the title,

Left: With or without the United States' business, cigars are still pulling in the big dollars in Cuba, and Fidel Castro is sitting pretty on a very profitable export.

Right: "I'll be back." Arnold Schwarzenegger smokes cigars between blockbuster films.

66

SMOKE

THE ENDURING EMBARGO

Cigars have peaked and waned in popularity, but for over twenty years, the United States embargo on trade with Cuba has endured, much to the disappointment and downright dismay of American lovers of fine Cuban cigars.

In 1961, United States President John F. Kennedy—himself an avid fan of H. Upmann Petit Coronas—ordered a total trade embargo with Cuba in retaliation for communist activities there. Rumor has it that before imposing the embargo, Kennedy sent Press Secretary Pierre Salinger on a Washington D.C. treasure hunt in search of as many of the President's beloved Cuban cigars as he could round up. While the embargo left countless Americans Cuban-less, Kennedy enjoyed a fine supply, proving that it is good, after all, to be President.

Meanwhile, back in Cuba, prime tobacco continues to be grown and fashioned into the finest cigars the world. Even after over thirty-five years without the United States' business, cigars remain Cuba's fourth biggest export, and play a vital role in the Cuban economy. But ironically, the best Cuban cigars aren't available in Cuba—they're earmarked for export, leaving the second-rate smokes for the domestic market.

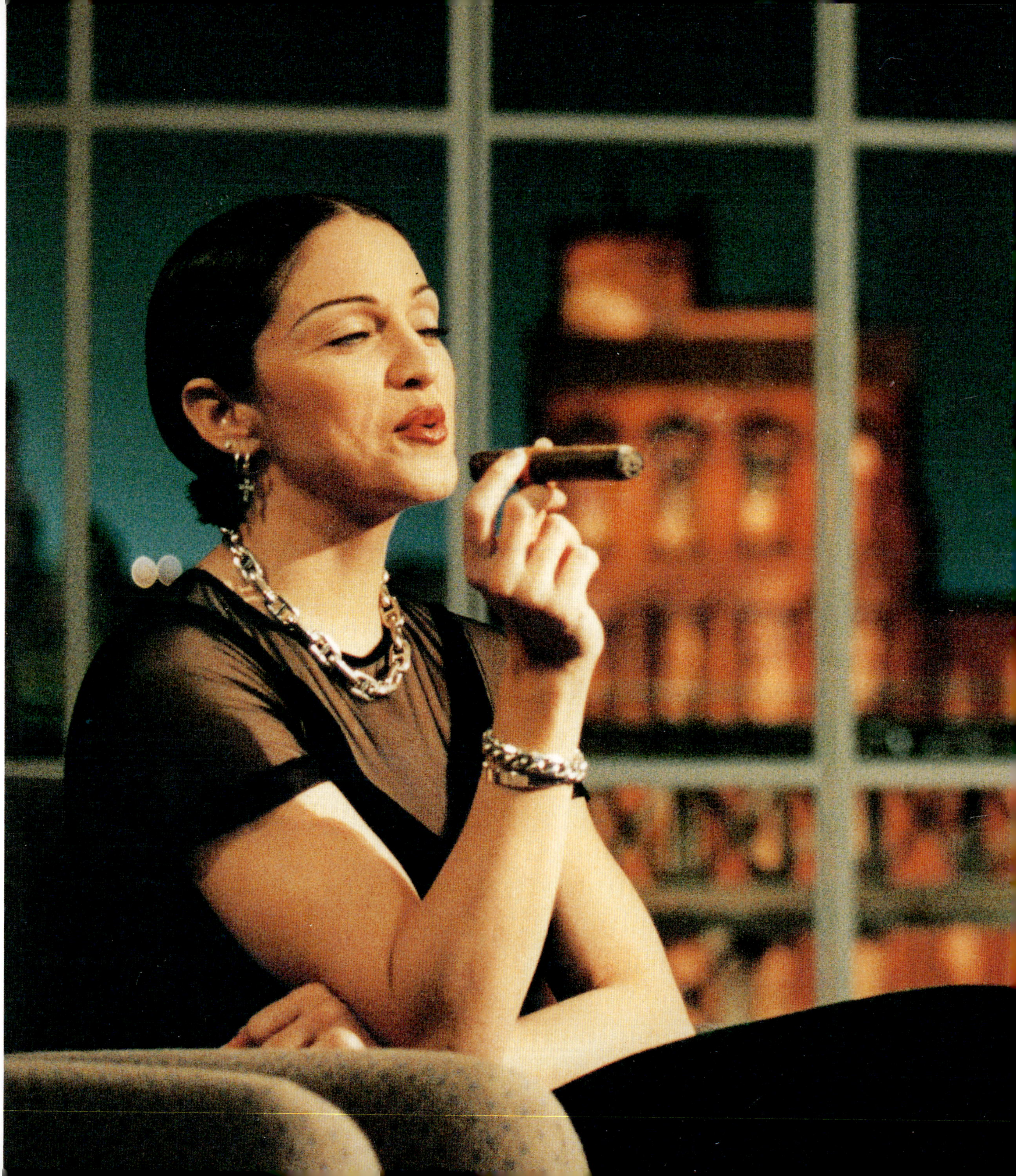

David Letterman is a well-known cigar lover, but even he was taken aback when Madonna appeared on his *Late Show* puffing on a cigar.

THE CIGAR

THE CIGAR

and it is only the most skilled who advance to rolling the finest cigars. Before a torcedor is commissioned to work on the puro, she or he—today there are more females than males performing the highly skilled work—must pass six very specific grades of excellence. A torcedor must be able to create, out of delicate and near flawless leaves, hundreds of cigars with identical shapes, dimensions, and colorings. They use only their hands and are, perhaps, the last of a dying breed of artisans.

Torcedors are not the only specialized members of the cigar production team. Before the leaf reaches their hands, a team of tobacco experts must choose those leaves from the drying racks at the barns. These teams are responsible for guaranteeing that their manufacturer's cigars will always be not only of the same quality, but of the identical flavor, body, and aroma as has been promised by the brand name. They must ensure this from cigar to cigar and from year to year, for a clientele whose tastes demand the best.

What Other Leaves?

While Cuba is reputed for producing the world's most popular cigars, there are other successful and appreciated cigar leaves from around the world. In Connecticut, a northeastern state in the United States, growers have perfected their leaves by combining local and imported seeds. These Connecticut leaves, with their smooth, fine veins,

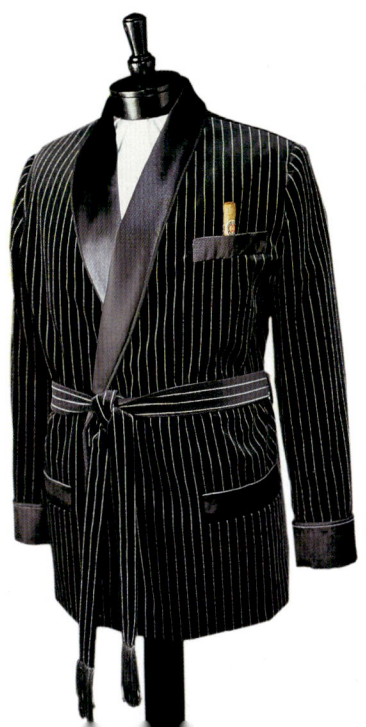

Opposite: **Cigar factory workers, Manicaragua, Cuba.**

Left: **Collage, Hugo Cloud, 1996.**

Below: **Savinelli chalk-stripe smoking jacket, by Lorraine Wardy Enterprises.**

SMOKE

The Cigar

"We all like people who do things, even if we only see their faces on a cigar-box lid."

—Willa Cather

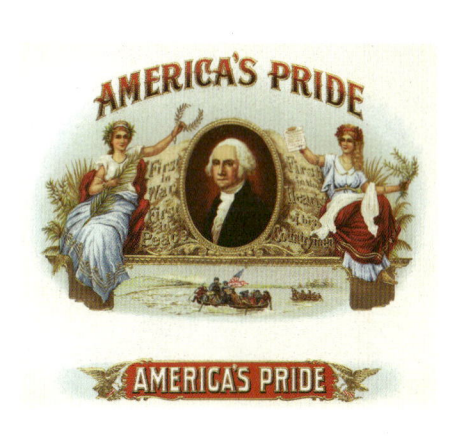

Above and right: **Cigar box labels, made in the United States with Yankee pride.**

Opposite: **Making cigar box labels, Havana Cuba.**

are often used as wrappers. In fact, cigars were very much a part of that area as early as 1770 when a merchant in nearby Pennsylvania opened a retail cigar store. Pennsylvania, for its part, is one of the largest producers of domestic dark and heavily bodied filler tobacco.

Sumatra, one of the many islands that make up the Southeast Asian country of Indonesia, also produces a fine quality leaf. It is considered the most elastic and attractive in color and, except for Cuban, is perhaps the finest wrapper available because of these qualities and its neutral flavor. Java, also an island of Indonesia, produces a similar quality leaf which is a bit darker and heavier than the Sumatran. Many of the cigars manufactured today use a combination of leaves to achieve the unique flavors needed for an increasingly diverse array of tastes in the cigar smokers' market.

The Final Package

If the leaves of the filler, binder, and wrapper are most important, perhaps the second most important aspects of

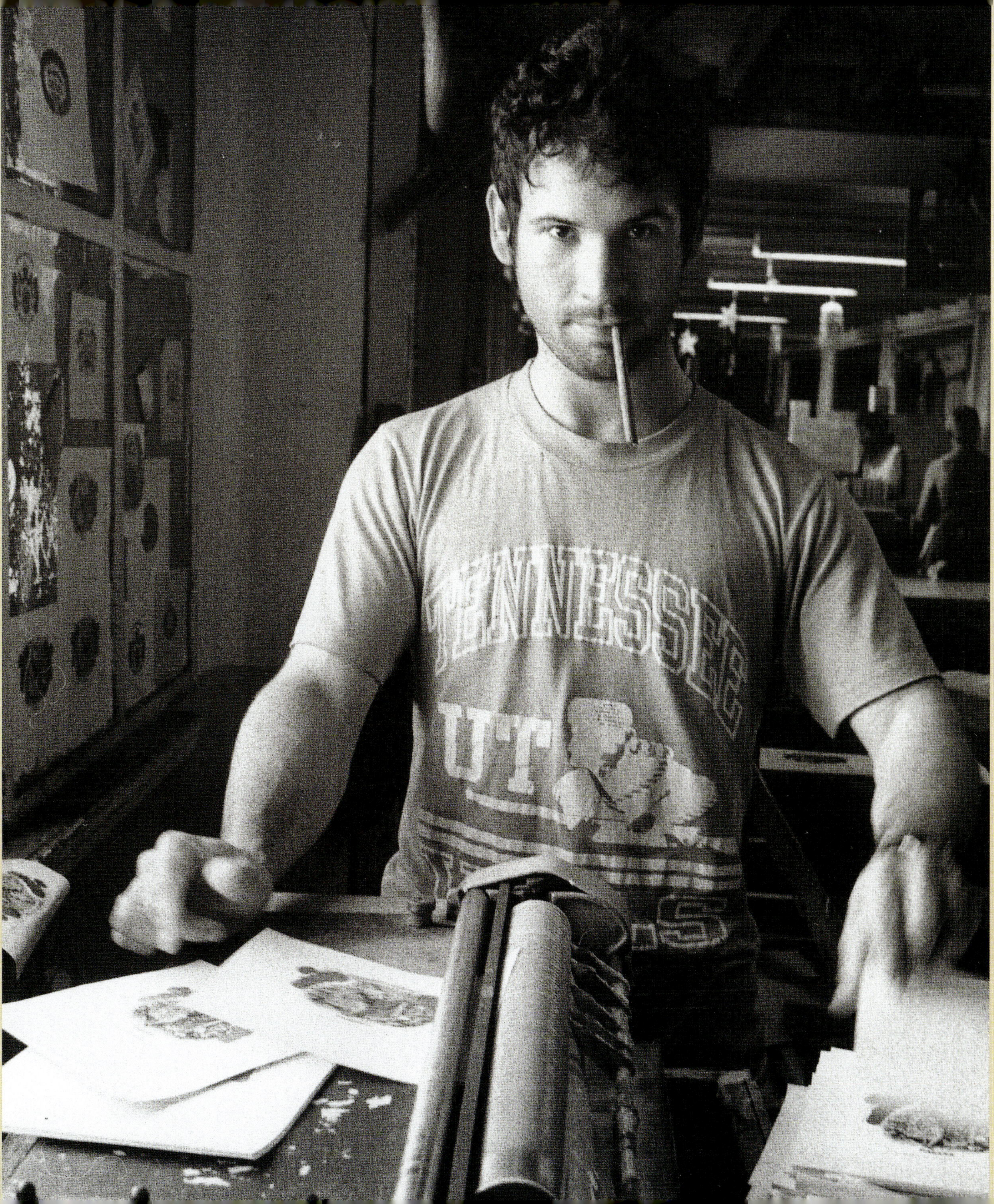
TENNESSEE
UT

Portrait of Stephane Mallarme, by Edouard Manet.

BETTER SMOKED AT HOME

A cigar is considered by some an indulgence best enjoyed in silence. One old tale illustrates the commitment some smokers have to that ideal:

An Englishman and A Frenchman were traveling together in a diligence, and both smoking. Monsieur did all in his power to draw his phlegmatic fellow passenger into conversation, but to no avail. At last, with a superabundance of politeness, Monsieur apologized for drawing his silent companion's attention to the fact that an ash from his cigar had fallen on his waistcoat, and that a spark was endangering his neckerchief.

The Englishman, now thoroughly aroused, exclaimed: "Why the devil can't you let me alone! Your coat-tail has been on fire for the last ten minutes, but I didn't bother you about it!"

Above: The return of the smoking lounge: Miami's Cuba Club.

cigar manufacturing are the boxes and bands. Cuban cigar manufacturers started to decorate their cigar boxes with beautiful and colorful pictures to set themselves apart. Companies around the world followed suit in an attempt to attract smokers with images of cigar smoking Nirvana and the pleasures of the good life.

Today there is a vast and impressive history of pictures and stories told on both the inside and outside of cigar boxes and around the tips of cigars. For the true Havana cigar, the green Cuban government tobacco seal is required on all exports. If a box does not have this seal, the cigars are not true Havanas. Many members of royalty, presidents,

VAUTIER
CESAR
CENDRE BLANCHE

Right: **Cigar with guillotine cutter by Solingen.**

Opposite: **Red Auerbach, former coach and president of the Boston Celtics smokes a signature cigar.**

THE CIGAR

and prime ministers have had their images put on the bands of fine cigars.

Cigars Sweep
A Curious Contemporary Culture

Today cigars and cigar lovers are enjoying an international resurgence in the popularity of their pleasure. After a steady but silent few decades, the nineties have made cigar smoking chic once again. Cigar bars are cropping up throughout the world's cities, and women, especially, are taking advantage of the trend. Restaurants are sponsoring "cigar nights," and places where cigars were forbidden before and cigarettes allowed, are now turning in favor of the older and longer smoke.

This latest cigar-smoking trend may indicate a popular attitude closer to that of modern tobacco connoisseurs and precisely the opposite attitude prevalent

> "This vice brings in 100 million francs each year. I will certainly forbid it at once—as soon as you can name a virtue that brings in as much revenue."
>
> —NAPOLEON BONAPARTE

The Cigar

Once A Lover, Always A Lover

Those who love their smoke, love it well.

When life was all a summer day,
And I was under twenty,
Three loves were scattered in my way—
And three at once are plenty.
Three hearts, if offered with a grace,
One thinks not of refusing.
The task in this especial case
Was only that of choosing.
I knew not which to make my pet—
My pipe, cigar, or cigarette.

—Henry S. Leigh

Above, right: A smoker's paradise: Arnold's Tobacco shop in midtown Manhattan.

Opposite: Perhaps Lucille Ball had a weakness for all things Cuban. Here, she puffs on a cigar in Lover Come Back, *1947.*

during the initial explosion of cigarette use in the early 1900s. When the cigarette first gained its enormous popularity, it was touted as a quick fix for those "on the move" tobacco users in an era of rapid industrialization and shrinking leisure time. Now, as people seek to increase their leisure time, assert their opposition to a life of only work, and demand a slower pace, many are lighting cigars to prove just how long they can sit still.

It will take more than the fickle popular culture scene to establish cigars in the mainstream, especially in the face of today's powerful anti-smoking voice. True aficionados, however, may prefer to let the cigarette keep its stronghold on the tobacco debate, agreeing that a cigar is most enjoyable at home.

"I toiled after it, sir, as some men toil after virtue."

—Charles Lamb, when asked by his doctor how he had acquired his power of smoking at such a rate

Smoke

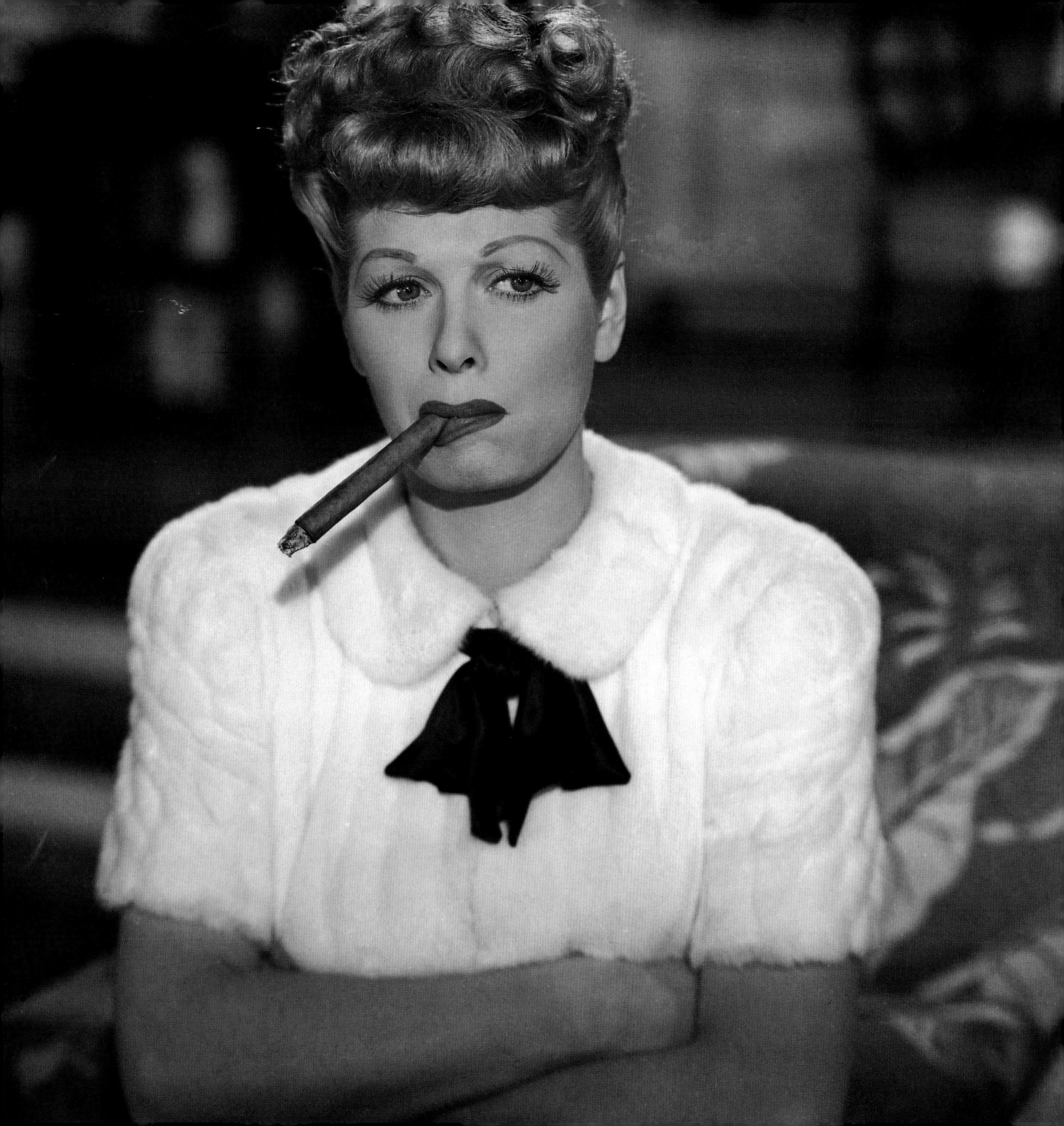

Right: The smoke makes the man... Gary Cooper, 1934.

CHAPTER FOUR

THE CIGARETTE

The Littlest Giant

Looking at the rise and fall of other tobacco uses throughout history, one might have predicted a similar fate for the cigarette. When the Crimean War ended, who could have known that the diminutive tobacco sticks brought home by European soldiers would create one of the world's largest industries?

"The man who smokes, thinks like a sage and acts like a Samaritan."

—Edward Bulwer Lytton

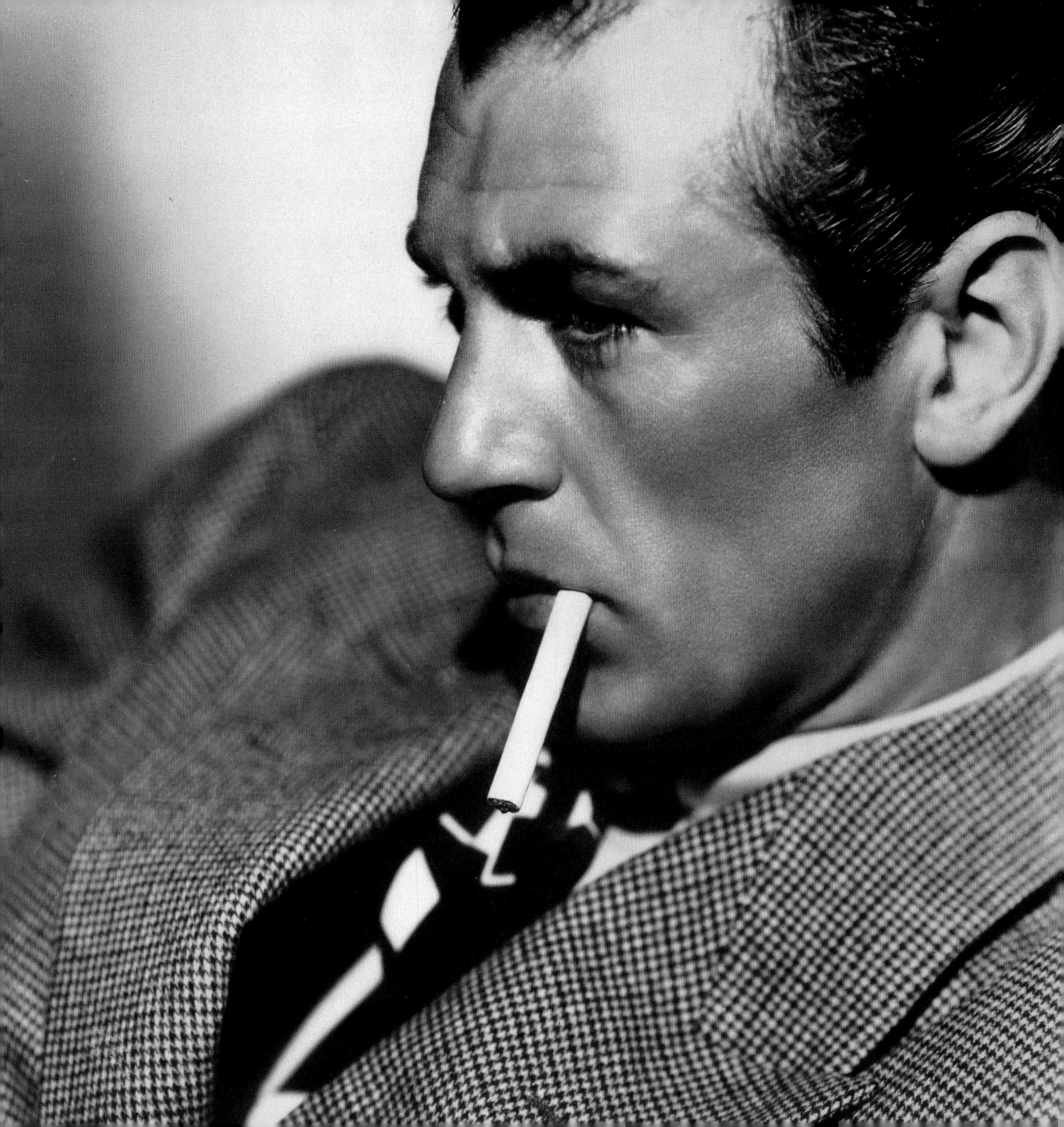

THE CIGARETTE

TRULY DEDICATED COLLECTORS

While cigar boxes and bands have long been hailed for their intricate and eye-striking pictures, many collectors throughout the world have set their sights on the innumerable cigarette packages created over the last century. Manufacturers put millions of dollars to make their brand easily recognizable. With marketing schemes ranging from subliminal pictures to personalized packs, it is no wonder that a camp of dedicated collectors and packaging experts has emerged from the fanfare. These cigarette packs chronicle the many phases of printing, lithography, and graphic design in the packaging industry. One Danish collection has over 53,000 packs saved over many years from 210 countries by a lawyer, Niels Ventegodt. The many package images are utterly diverse including pagodas from Thailand, a war scene from Great Britain, the Mona Lisa on a Danish brand, and a couple dancing the Tango on a pack from Malta.

THE CIGARETTE

For many reasons, the cigarette took hold of the world and has been virtually unceasing in its expansion. Even today, while a fiery debate rages in America, and Europeans and Australians become increasingly concerned about the repercussions of smoking, the market for tobacco—and specifically cigarettes—is still growing in Central and South America, Africa, Eastern Europe, and Asia.

Once the quality of United States-grown tobacco earned its reputation, savvy entrepreneurs went on a hunt for untapped opportunity. James Duke, son of a North Carolina tobacco man, saw difficult times for his family's business in the face of furious competition. In capitulation, he decided to put his efforts into pre-rolled (as opposed to the standard rolling-by-hand method of the time) cigarette manufacturing and gutsy marketing. He pounced on new photo and print technology to advertise his product and kept tight tabs on competitor activity. Within twenty years, Duke had established himself at the

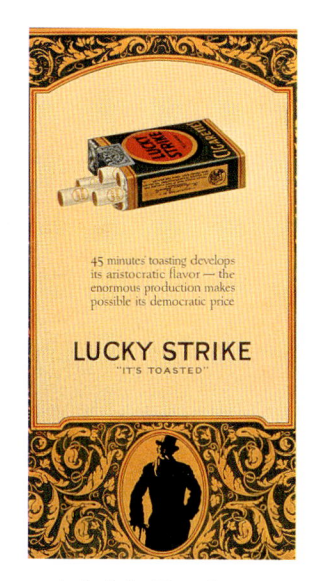

Left: John Travolta was nominated for an oscar for 1994's *Pulp Fiction*. As sentimental hitman Vincent Vega, Travolta made his way through the film shooting off guns and shooting up heroin, and smoking lots of butts.

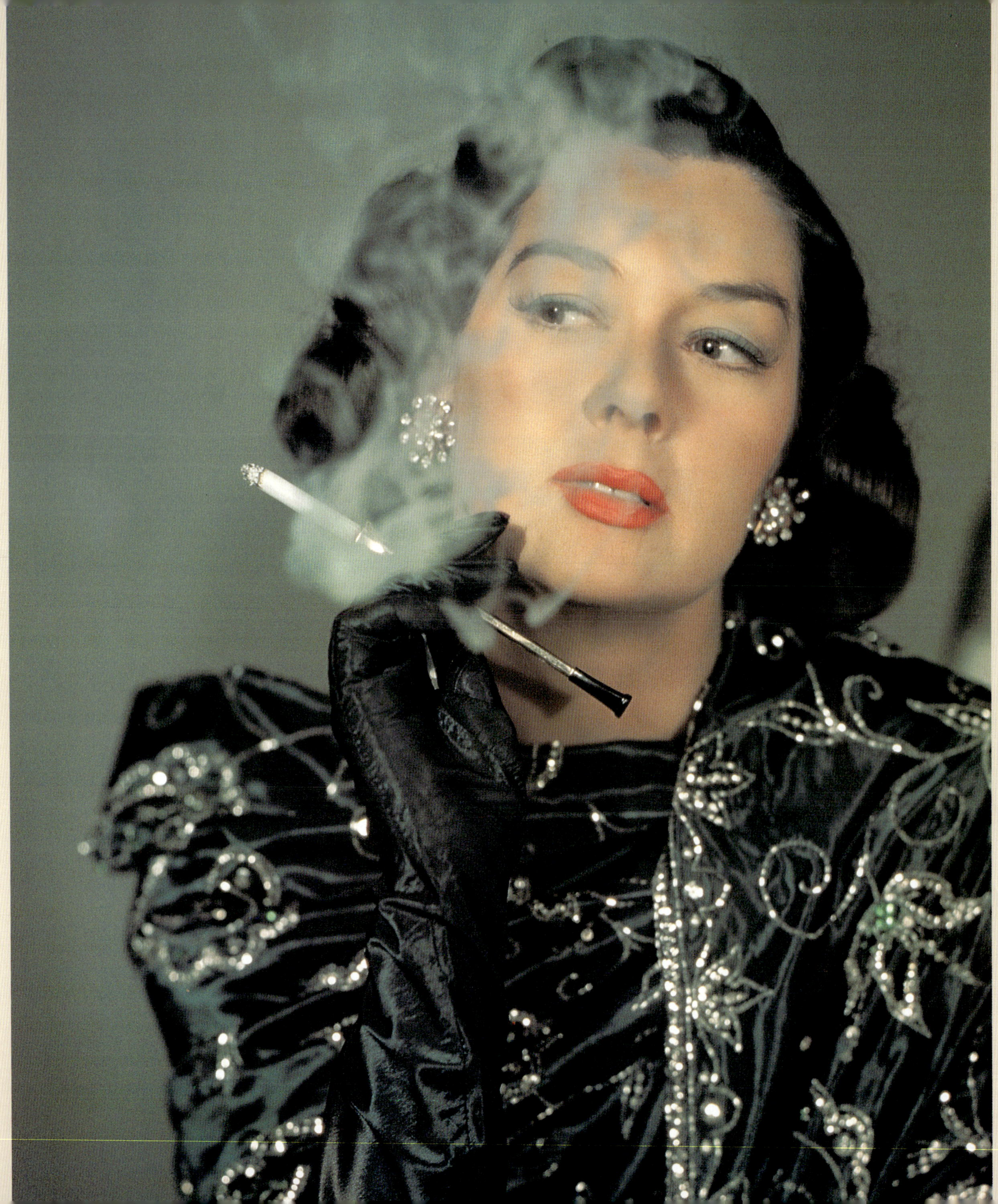

THE CIGARETTE

pinnacle of the global cigarette market. Duke University, one of the nation's finest schools, was built by the Duke family with their tobacco fortune.

Why All The Noise?

A combination of the cigarette's inherent appeal—a quick fix of stimulant for the "movers and shakers" in society, a rebellion against the conservative social restraints of the time, and a (still misinformed) notion of certain benefits of smoking—along with the industry's increasingly ubiquitous advertisements, played into the hands of the masses.

In 1914, advertising campaigns went national and the cigarette took off to become the world's most popular form of tobacco use. Camel brand cigarettes introduced itself over four days in the *Saturday Evening Post*, first ambiguously and finally, with a blaring notice that screamed, "Camel cigarettes are here!!" Other manufacturers started to include gifts for the purchaser in every pack.

Duke included the now famous cigarette cards in his packages. The cards served as both an enticement and a pack stiffener to protect the cigarettes.

Opposite: What starts out as merely cool becomes perfectly glamorous when you add a six-inch cigarette holder. Rosalind Russell, undated.

Above: This cigarette rolling machine can turn out 800 cigarettes in a minute. Here, a woman grabs a giant handful of the finished product.

Left: Advertising just ain't what it used to be. Advertisement for Marshall's cigarettes, late nineteenth century.

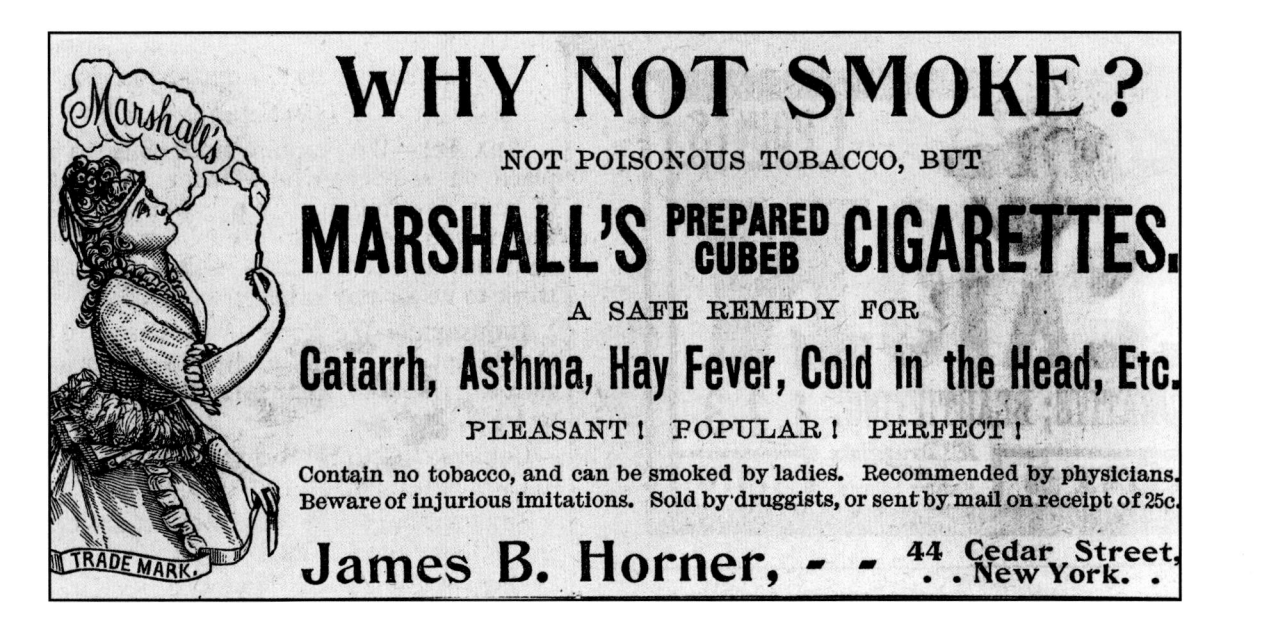

THE CIGARETTE

THE GLAMOUR OF IT ALL

Japan has a solid base in the tobacco market with their popular and internationally recognized Mild Seven cigarettes. Now a cosmetics firm has found another angle from which to reach smoking consumers. Shiseido, a well-known cosmetics line, has produced and is now marketing hair care products that help to combat the smell of tobacco that tends to remain in women's hair after smoking or socializing in a smoky atmosphere. Reminiscent of the velvet caps and smoking jackets that were expressly designed to absorb cigar odors in the 1800s, Shiseido is taking an old idea and targeting a new female audience.

Below, left: Irene Bordoni, a star of the New York stage in the 1930s, returned from a European vacation with a pretzel cigarette holder, giving new meaning to the concept of curls of smoke.

Right: Though they don't turn up too often nowadays, the cigarette was once a popular prop for the Hollywood glamour photo. Corrine Calvert, 1948.

With the majority of the market still male, manufacturers found that images of scantily clad women made their particular variety of smoke far more appealing. When speaking of his childhood memories, American novelist and essayist Henry Miller once said: "That first burlesque show I shall never forget. From the moment the curtain rose I was trembling with excitement. Until then I had never seen a woman undressed in public. I had seen pictures of women in tights from childhood, thanks to Sweet Caporal cigarettes, in every package of which there used to be a little playing card featuring one of the famous soubrettes of the day."

"A cigarette is the perfect type of a perfect pleasure. It is exquisite, and it leaves one unsatisfied. What more can one want?"

—OSCAR WILDE

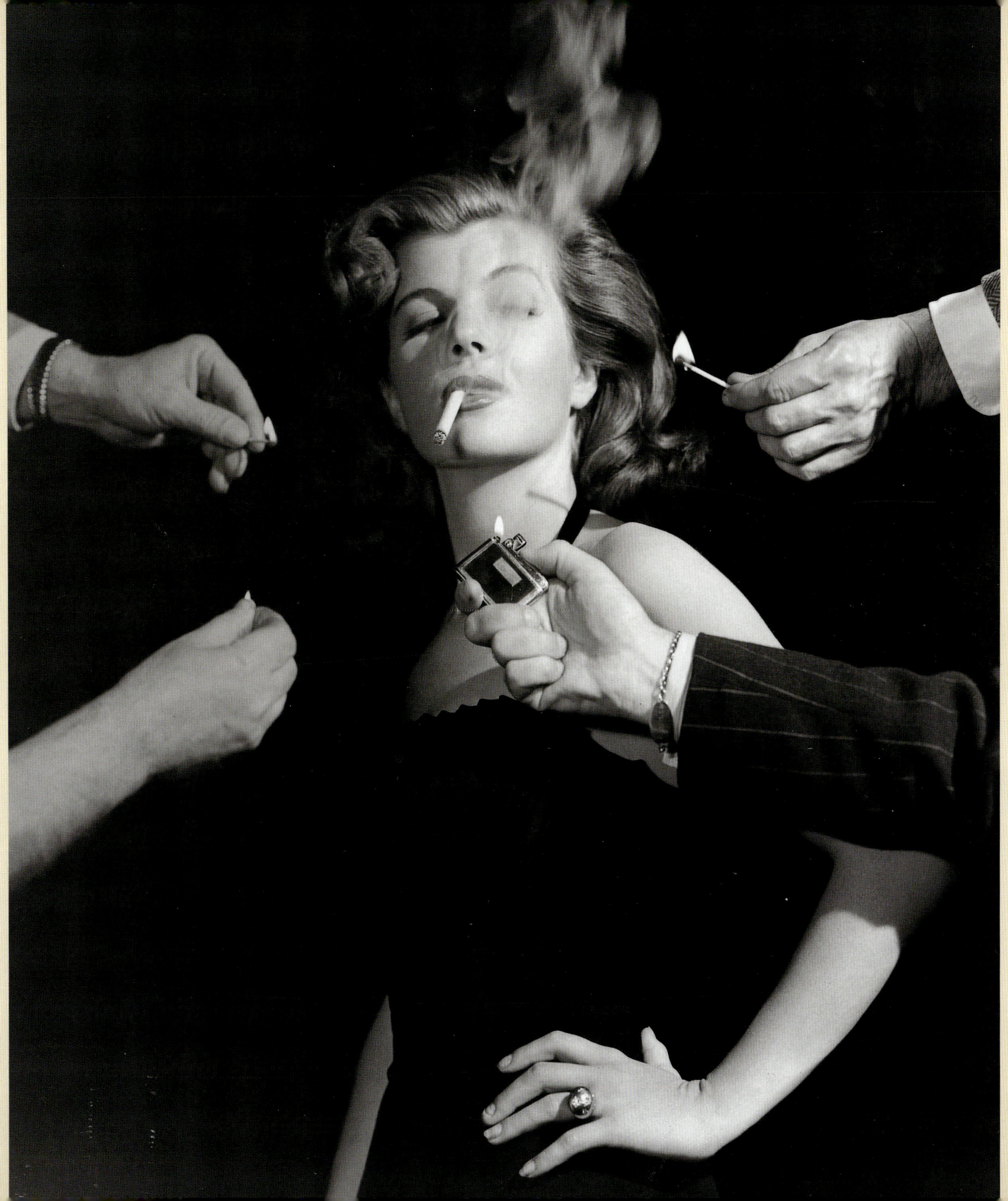

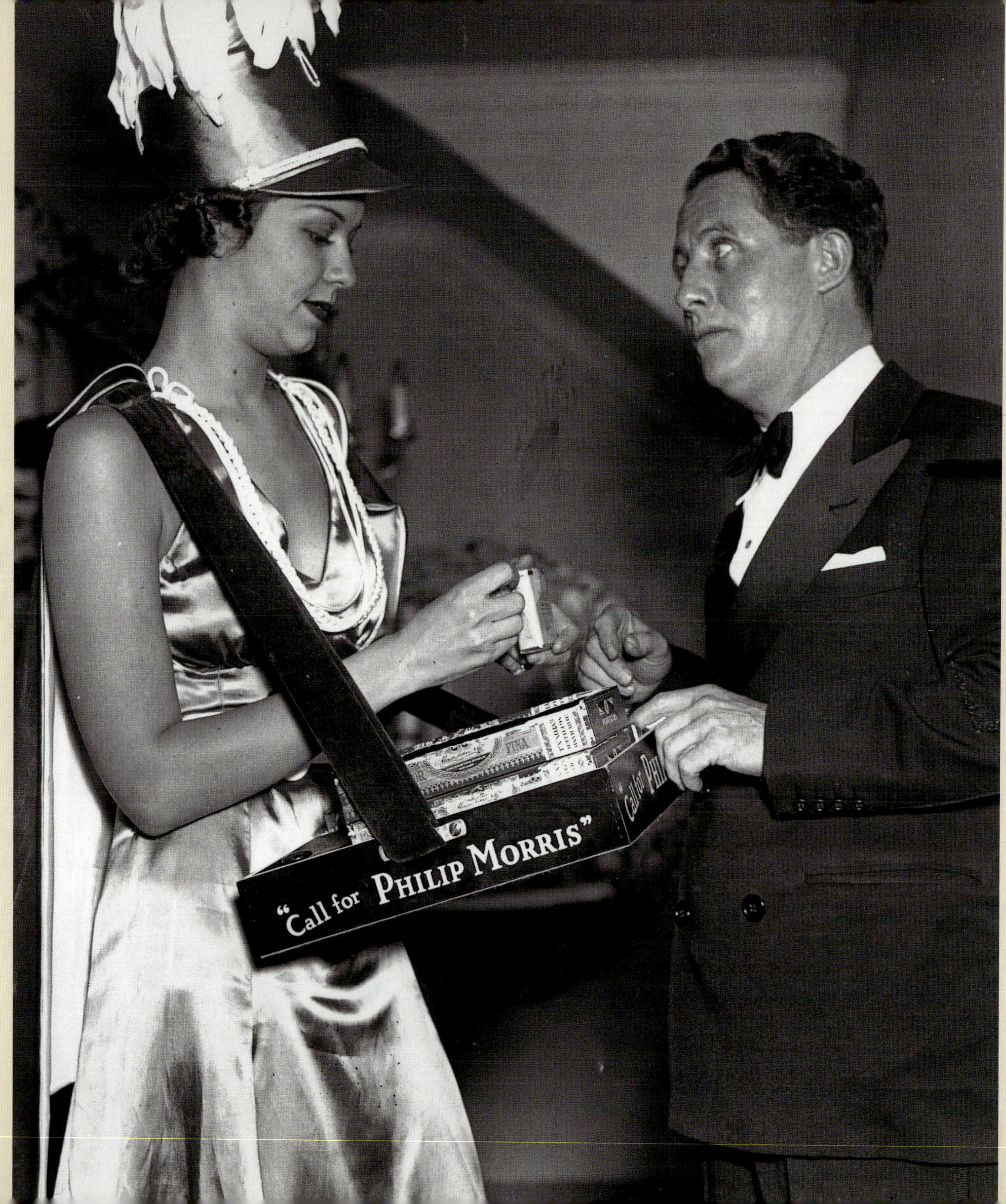
"Call for PHILIP MORRIS"

THE CIGARETTE

Left: Once upon a time, not only could you smoke just about everywhere, you hardly had to leave your seat to buy a pack. Movie comedian Bert Wheeler stocks up at the Cotton Club in New York City, 1936.

Top right: The Hollywood tough-guy tradition continues: Johnny Depp lights up at the Cannes Film Festival, 1995.

Below right: European theater ad, 1920.

Women: A Marketing Bullseye

It was not long before the marketing minds of the time broke new ground by targeting a growing female clientele. During the early 1920s, cigarette companies took advantage of the suffragettes by loosely tying the fight for voter rights and equality to the social right of smoking. A psychiatrist hired by the American Tobacco Company asserted, "Some women regard cigarettes as symbols of freedom. . . . Cigarettes, which are equated with men, become torches of freedom." The touting of equality by one company was ironically juxtaposed by Lucky Strike's infamous campaign toward women; smoke cigarettes instead of eating candy to lose weight and be attractive.

CELEBRITY CIGARETTE USE

Cigarette companies sponsor much of what we see on television and always have. Before advertising here was banned in 1971, television personalities often included tobacco smoking in their shows.

Famous people were—and still are but to a lesser extent—veritable walking billboards for the folly. Marlene Dietrich, Lilli Palmer, Rock Hudson, Ronald Reagan, Gregory Peck, Jane Wyman, Bob Hope, Jack Webb, Laurel and Hardy, Eva Gabor, William Holden, Henry Fonda, Tony Curtis, and Fred Astaire have all touted the "mildness" and fine flavor of America's cigarettes. Foreign diplomats, literary figures, cartoon characters, and even animals had a role in the gimmickry of the age.

Musicians, too, have always been notorious for the clouds of swirling smoke around their sound, and the advent of rock and roll further entrenched the cigarette into that venue. With rock and roll rebellion came the ever-present cigarette, and icons like Keith Richards of the Rolling Stones—known the world over for his low slung guitar and ever-present cigarette—made visual images that are as transcendent as the music they play. Jazz musicians also fit their smoky music into smoky clubs, often playing with a cigarette gripped between their teeth as well.

THE CIGARETTE

For thy sake,

tobacco, I

Would do any

-thing but die.

—CHARLES LAMB

Other companies tried a more "aesthetic" approach. The Player Tobacco Company included "silks" for female smokers. These tiny pieces of fabric were advertised as a virtual house-decorating genius, "the satin inserts may be stitched (herringbone style) on a piece of silk, satin, or other material. As indicated, some simple embroidery adds greatly to the attractiveness of the finished article. Other things which can be made from the insert are screens, bedspreads, lamp shades, sewing bags, hat bands, portieres, pin cushions, doilies, table centers, masquerade dresses, belts, bands for the hair, kimonos, pillow tops, tiers, piano drapes, tablecloths, doll's dresses, teapot cosies, egg cosies, mantel drapes, comforters, handkerchief bags, sideboard covers, dresser covers, covers for chairs, parasols, etc. A colored sheet showing some of the uses to

When green dye became scarce during World War II, Lucky Strikes cigarettes changed from their then-familiar green packaging to white boxes to help the war effort. The brand shipped tons of smokes over to the boys on the front lines, like these soldiers in Okinawa in 1945.

"You ask me what we need to win the war? I answer tobacco as much as bullets."

—GENERAL JOHN J. PERSHING

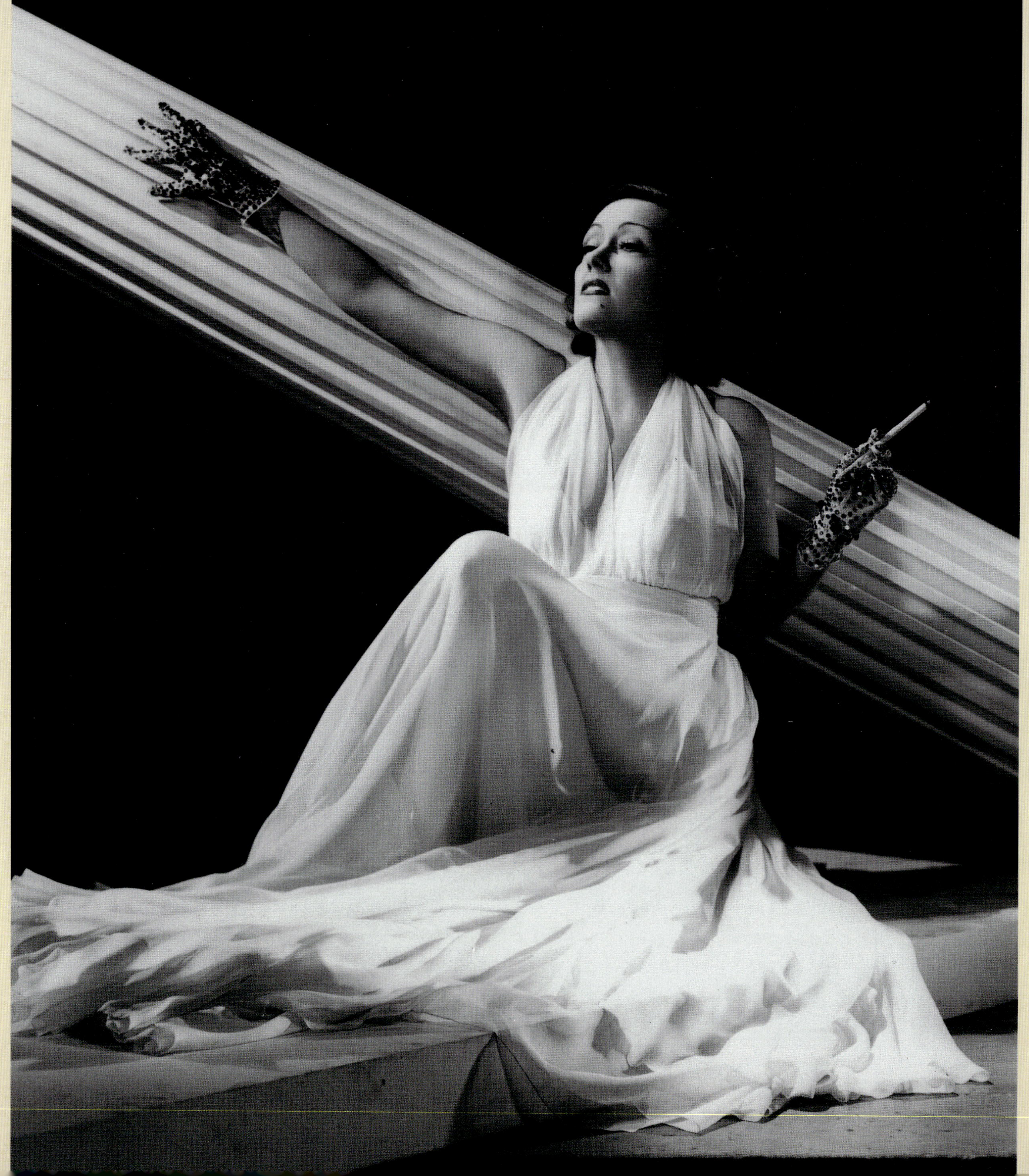

THE CIGARETTE

which these inserts may be put will be mailed to you post free on receipt of your name and postal address." The marketing was smashingly successful in both tempering social abhorrence of female smoking and attracting new smokers. By 1929, American women were smoking nearly fourteen billion cigarettes per year.

In the Hollywood of old, cigarettes signaled sophistication and glamour; in the nineties, they tend to suggest a kind of rugged masculinity. *Left:* **Gloria Swanson, star of the silver screen, 1940.** *Right:* **Tim Roth, tough guy of the independent film scene, 1996.**

The Sights And Sounds Of Smoke

Radio and cinema, supported by the tobacco industry, further propelled the cigarette on its racy course. The 1940s provided the cigarette's last decade of unscathed popularity. Big bandleaders syncopated their tapping feet to the tune of their puffing "fag" while some of the greatest movie stars in American history sent clouds of smoky sensuality up and across the screen. Cigarette-brand icons, like Willie the Penguin for Kool Cigarettes, were of the most easily recognized symbols around. Ashtrays, end tables, lighters, jackets and robes, make-up compacts, key chains, cigarette holders, magazine racks, walking sticks, and even champagne bottles were all made to accommodate the glamorous life of the smoker. If a brand name could be stuck on it, you could be sure it was used to sell cigarettes.

THOSE SPICY CLOVES

There's the smell of a cigarette, then there's the smell of a clove cigarette. While cigarette smoking seems as common as breathing—much to the dismay of those who don't relish breathing in cigarette smoke—there's a certain, sweet and spicy smoke filling the air some of the more sophisticated pubs and lounges in major cities and college towns. The distinctive, pleasant smell of cloves, combined with the exotic appeal of something imported, expensive, and sophisticated make clove cigarettes a popular accessory for hip intellectuals and smokers looking to stand out from the nicotine-drenched masses.

Clove cigarettes, or kreteks, as they are properly called, are a blend of fine tobacco and clove spices, and have enjoyed a quiet but substantial niche market for smokers looking for an alternative to standard cigarettes. Imported from Indonesia, where they originated, cloves contain a higher quality tobacco than standard cigarettes, are more densely packed, and contain no fillers, which makes them burn more slowly. And while the health risks of cloves are as real as those of regular cigarettes, fans still consider them a more natural and—mistakenly—healthy alternative. In a word obsessed with alterna-culture, these pungent yet pleasant-smelling smokes, wrapped in distinctive, dark paper, make a bold and sophisticated fashion statement.

NO LONGER A POPULAR PASTIME

"Brother G.," said one clergyman to another, "is it possible you smoke tobacco? Pray, give up the unseemly practice. It is alike unclerical and uncleanly. Tobacco! Why, my dear brother, even a pig would not smoke so vile a weed!"

Brother G. delivered a mild outpouring of tobacco fumes, and then as mildly said, "I suppose, Brother C., you don't smoke?"

"No, indeed!" exclaimed his friend, with virtuous horror.

After another puff or two, then Brother G., who prefers the Socratic method of argument, rejoined, "Then, dear brother, which is more like the pig—you or I?"

Left: **In the early sixties, if the Hollywood Rat Pack was doing it, it must have been cool. Dean Martin, Sammy Davis, Jr., and Frank Sinatra (left to right) have a smoke as they relax offstage at Carnegie Hall, 1961.**

Far left: Self-Portrait with Cigarette, by Edvard Munch, 1895.

The Cigarette

Above: Lighting up when the sun goes down.

Right: Yes, some of the ladies in Hollywood, like Susan Sarandon, still do smoke.

World War II also played an enormous role in the cigarette's rank. Just as George Washington had declared during the American Revolution almost two hundred years before, the government and high ranking military officials deemed cigarettes invaluable in maintaining soldier esteem. General Douglas MacArthur told one corporate contributor, "The entire amount [of donation money] should be used to buy American cigarettes which, of all personal comforts, are the most difficult to obtain here." Because the dye used in Lucky Strike's green packaging became a needed commodity during the war, the company put out a new (and still used) white package and hailed itself as a great benefactor to the national effort.

Harris Lewine, another American author reminisces upon smoking in high school, "Those long lunch hours became a gauntlet of all the smoker's tricks, affectations, and pleasures. There were: smoke rings without inhaling (the correct way); smoke rings with inhaling, the rings being made out of the exhale (difficult); French inhaling (which I practiced but couldn't master); lighting two cigarettes at once—one for your girl, like Paul Henried for Bette Davis; the different methods for lighting a match against the wind; the one-handed strike, light, and match-discard; the thumbnail-stick match light; the correct way to flip a 'butt'; how to talk with a cigarette in your mouth without choking; the 'Russian held' smoking style, the butt behind the ear, either as a 'save' or a make ready for the next smoking break. . . ."

A Fad Gone Mad

The joy ride of the cigarette slowed down, however, as the middle of the century came around the corner. The 1950s marked opposition's first significant step toward curtailing the rage.

98   SMOKE

THE CIGARETTE

"A Coke at snack time,

a drink before dinner,

a cup of coffee after dinner,

a cigarette with the

coffee—very relaxing.

Four shots of drugs.

Domesticated ones."

—ADAM SMITH

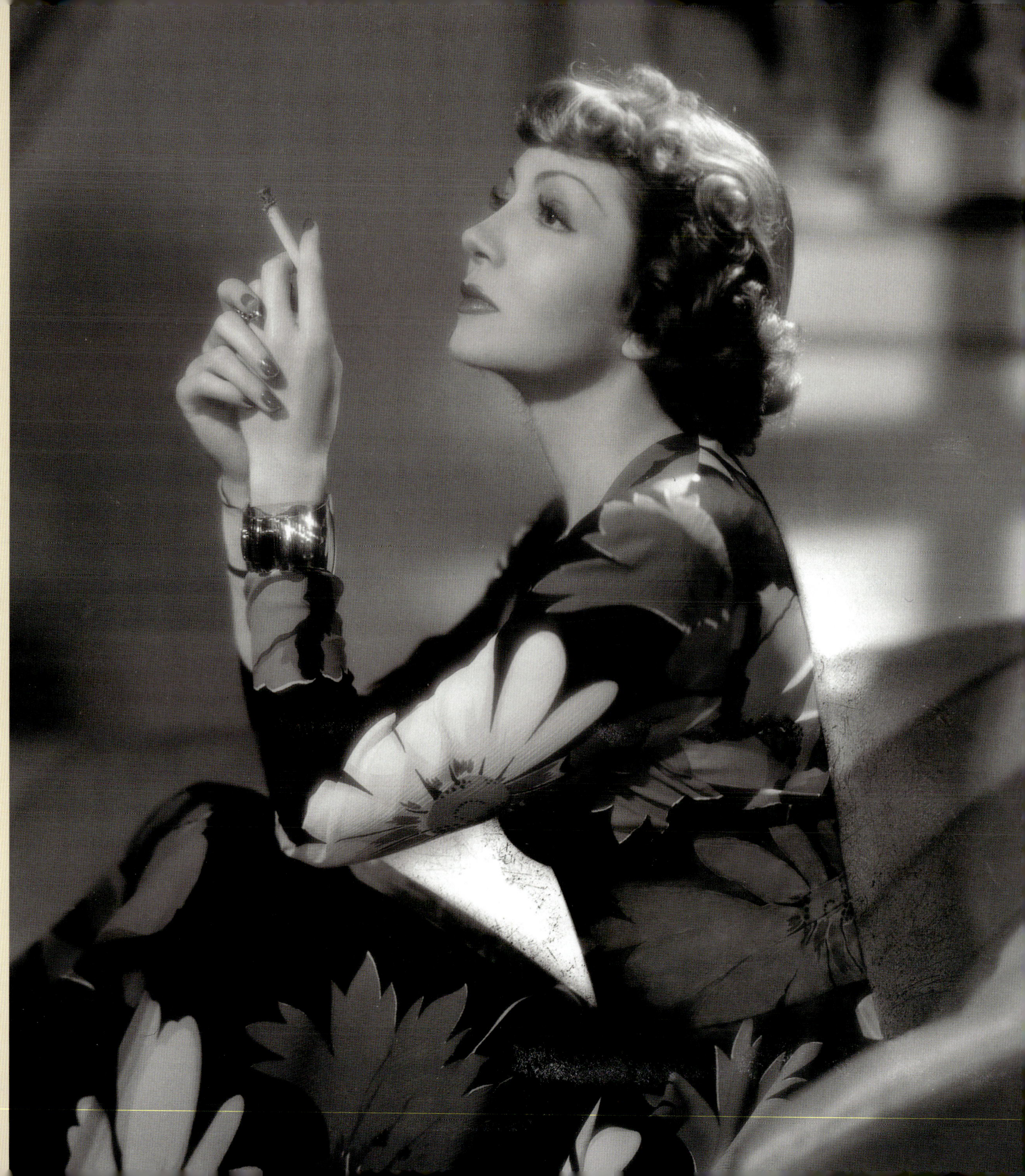

THE CIGARETTE

After the *Journal of the American Medical Association* made a definitive statement linking smoking to lung cancer, other more mainstream periodicals questioned the craze as well.

Suddenly, smokers were faced with more than a fashion decision and the results—a marked and steady decrease in the number of smokers—rallied the tobacco industry to not only a heightened advertising campaign, but to a more defensive position in their trade. They created their own investigative team, the Tobacco Institute Research Committee, to contest the findings of the anti-tobacconists. Certain brands were gaining market share but the shifts within the industry, with one brand overtaking another, or new products being introduced, had little effect on the fury surrounding the product.

In Europe, too, public cognizance was rising. The British Royal College of Physicians of London expressed concern about tobacco smoking in 1962. Thereafter in 1966, the Surgeon General of the United States required all packs to be stamped with warning labels. In 1971, television advertisements—heretofore a major promotional

"Food is now available in such unpleasant forms that one frequently finds smoking between courses to ba an aid to the digestion."

—FRAN LEBOWITZ

Opposite: **Claudette Colbert, 1938.**

Above: **Ex-cigarette spokestoon Joe Camel, for Camel cigarettes, 1997.**

"PEDIGREE, n.

The known part of the

route from an

arboreal ancestor with a

swim bladder to an

urban descendant with a

cigarette."

—AMBROSE BIERCE,
The Devil's Dictionary

THE CIGARETTE

runway—were prohibited. Celebrities, too, took up the fight against the once-loved smoke, disenchanting some and earning points with others. Throughout the 1970s, America's tobacco consumption dropped by more than ten percent.

Going Out With A Boom

Ironically, it was at this time that Marlboro created one of the most recognizable advertising campaigns in the world: the Marlboro Man. Filtered cigarettes had not yet penetrated the "masculine" market and Marlboro was a very weak competitor in the game. Prior to its upcoming sweep of the world's smoking majority, Marlboro had served a predominantly female clientele.

By appealing to males with images of the strapping stud they'd all hoped to be, Marlboro sales soared. The red-boxed cigarettes became—and still are—the world's number one-selling brand. While the Marlboro man is a fading icon in Western societies, one can hardly travel ten steps in the company's international market territory without seeing "Marlboro country" on a billboard above head.

Current Standings

It seems that cigarette consumption in the United States has now stabilized while the international market

Opposite: Ad for Job cigarettes, 1912.

Left: Coffee, cigarette, and the newspaper, or as they call it in **New York City,** breakfast.

SMOKE

THE CIGARETTE

THE DOPE ON POT

Legal or illegal, it seems that marijuana is here to stay. Smoked in pipes, cigarettes, bowls, water bongs, or baked in brownies, the flowering tops of the hemp plant have been used for achieving euphoria since ancient times. Evidence of its use in China dates back to 2737 B.C.; it spread to India, then North Africa, and reached Europe by 500 A.D.. It was a major crop in Colonial America, where it was grown as a source of fiber, and was extensively cultivated in the United States when Asian sources were cut off during World War II. It has produced its own unique vocabulary and lifestyle, which transcend economic and social class boundaries, and is as common in many circles as alcohol, if somewhat less obvious. Yet, cannabis and the drugs created from it—marjijuana and hashish—remain illegal in most parts of the world. Grown in backyards, basements, and dormitory closets, and traded quietly between acquaintances, the underground pot society thrives just beneath the surface of more accepted, legal vices like tobacco and alcohol, creating an unspoken bond that stretches around the world.

continues to increase. In 1996, American companies produced 760 billion cigarettes, a record for the industry. Tobacco farmers fear for the future of their crops in the face of stricter government regulation and look toward this extra-American market for their future livelihood and growth.

Today, the United States tobacco market includes almost fifty million smokers who dole out billions of dollars a year for their pastime. The six largest American tobacco companies, on the other hand, spend about $4 billion per year on advertising campaigns alone. It is estimated that the United States government spends approximately $97 billion per year on smoking-related healthcare, lost work hours, and educational programs. Phillip Morris is the market leader with 45 percent of the total domestic share and twelve percent of the international market, and RJR Nabisco, is a competitive second with their Camel brand cigarettes. In 1992, the United States exported $3.7 billion in tobacco (up 340 percent since 1975), nearly one-third of its entire crop. The cigarette holds the reigns on one of the swiftest industries around.

Right: **Jack Nicholson, with a smirk and a cigarette, can appear endearing and psychotic at the same time.**

Left: **A smoke of another kind:** *High Times* **magazine is dedicated to the celebration and legalization of marijuana.**

Right: **When you're running the country, you're entitled to a break once in a while. John F. Kennedy lights up at a Washington D.C. banquet, 1961.**

Left: **Asher Wertheimer, by John Singer Sargent, 1898.**

CHAPTER FIVE

THE TOBACCO MYSTIQUE

An International Phenomenon

Tobacco use spans the globe and predates written history. It's been transformed to fit tastes and times, and materials and lifestyles. Its power has paved international trade routes and pitted endless groups against one another in passionate debate. This seemingly innocuous herb and the multitude of ways we have developed its use, are of the greatest cultural phenomena in human history, and love it or hate it—it's here to stay.

"Now the only thing I miss about sex is the cigarette afterward."

—FLORENCE KING

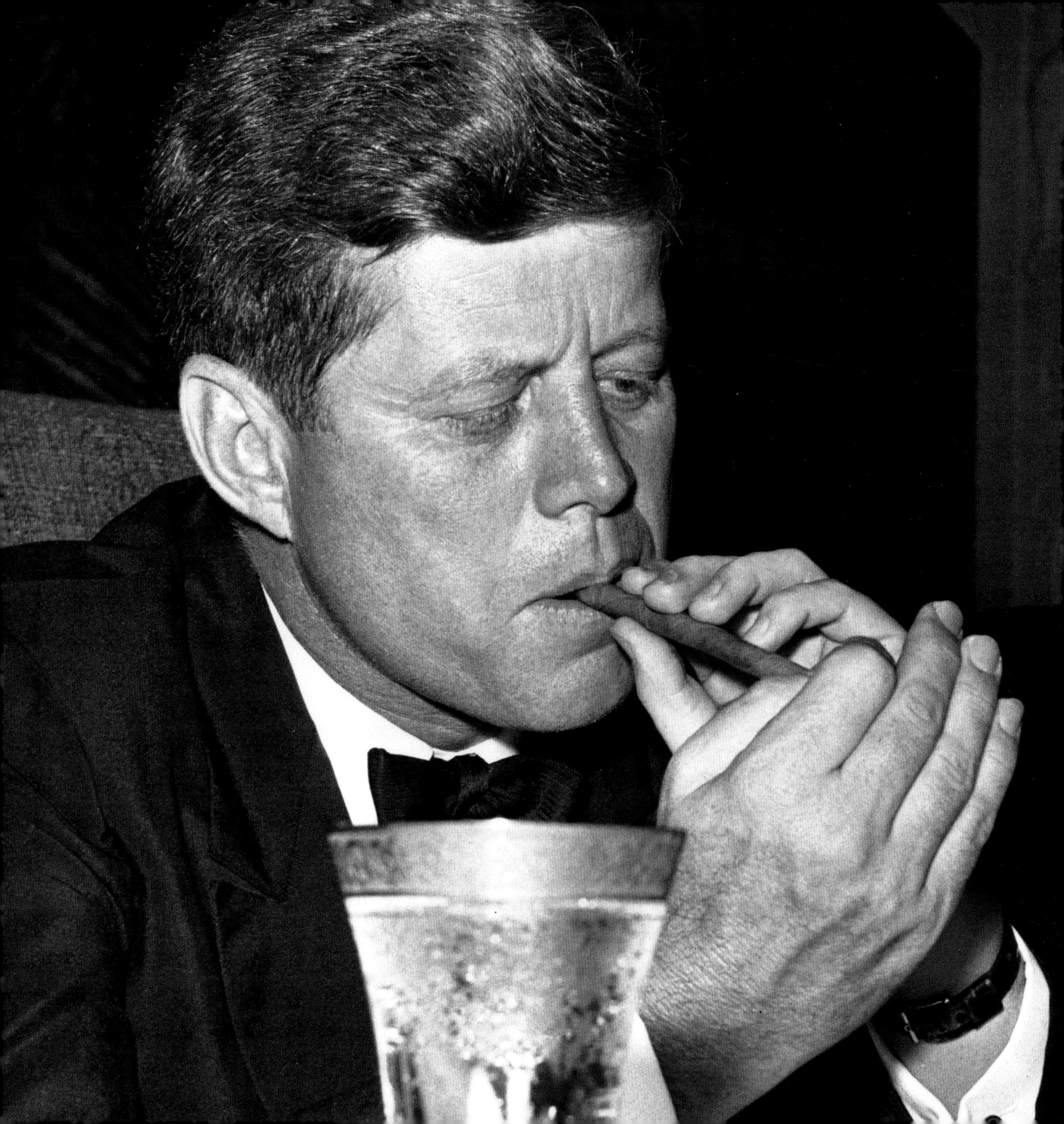

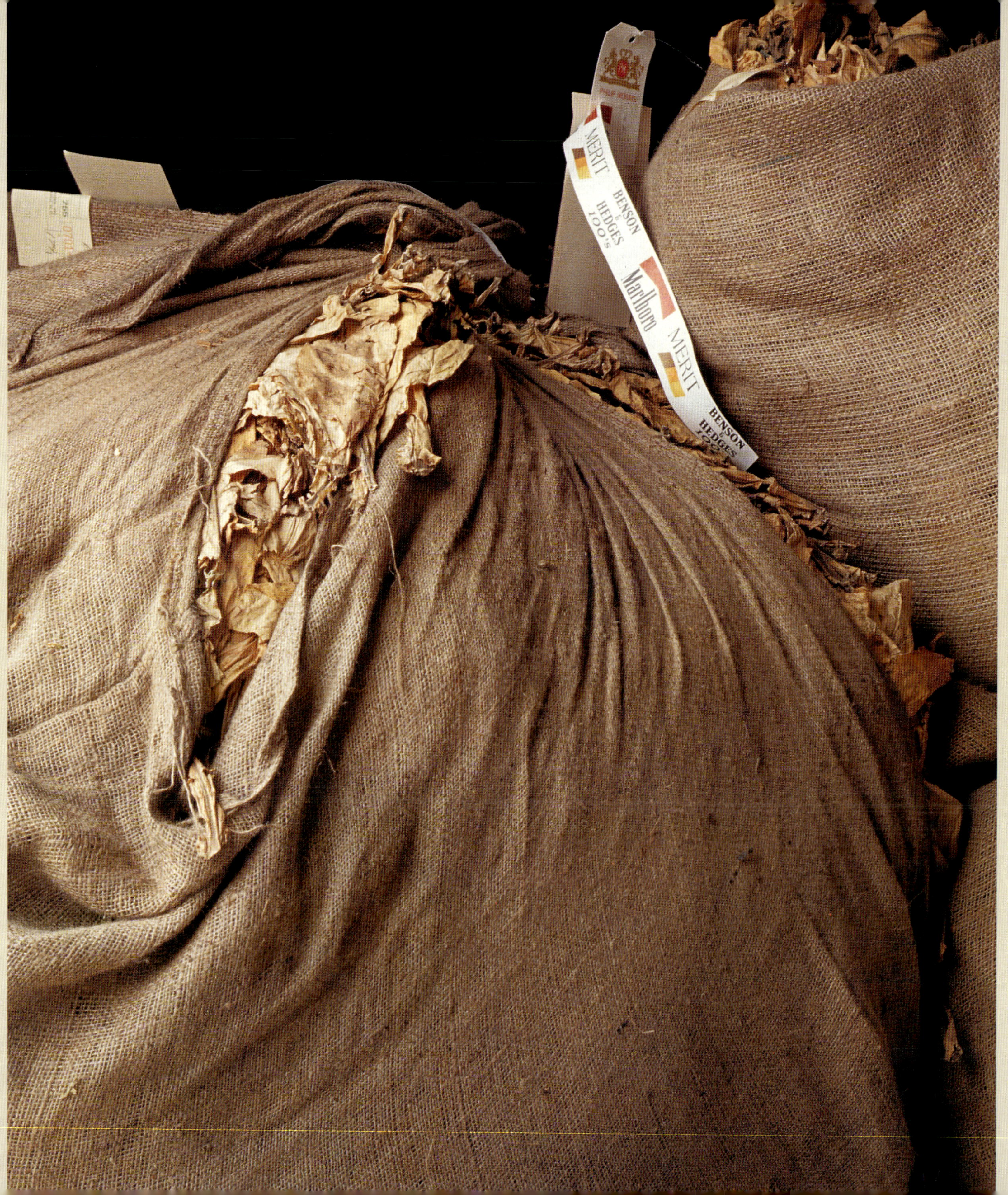
MERIT
BENSON & HEDGES 100's
Marlboro
MERIT
BENSON & HEDGES

THE TOBACCO MYSTIQUE

In its evolution, tobacco has crossed the lips and noses of a more diverse crowd than any of the world's finest foods or drinks. Each passing trend has armed tobacco with a new face and a new role to play. The sophisticates of the seventeenth century glamorized it. The merchants, sailors, and explorers of expansionist Europe toughened it. The cigarette card ladies gave it sex appeal and the movie stars of film's early years made it sensual. Great thinkers throughout history have bestowed upon it an air of intellectual necessity, while infamous madmen and women have sucked smoke through their trembling lips, exhaling an image of urgency we've come to recognize easily.

Fiction writers and film makers have taken all of these and sprinkled in a dash of corruption and depravity to further capitalize on all that smoking is. While the advertising industry has indubit-ably reveled in the market of smoking, it is the truly creative minds of past and popular culture that imbue the plant with personality. Because for now, cigarettes hold most strongly to the public's identification with tobacco. The following examples refer mainly to that form of smoke.

Smoking In Fiction

In his book, *Cigarettes Are Sublime* (Duke University Press, 1993), Robert Klein describes beautifully how authors can weave layers of imagery through their words by having a character smoke. "Extracted from its pack and smoked, the cigarette writes a poem, sings an aria, or choreographs a dance, narrating a story in signs that are written hieroglyphically in space and breath. . . ." He further discusses how thoroughly a cigarette identifies a character by comparing its use in literature to the word "I," making an introduction through a familiar gesture.

It is no wonder then, that great writers from George Sand to Cormac McCarthy have included smoking scenes in their journals and fictional works. How better can a writer create an image

Opposite: **Cured tobacco awaits shipment, Dillon, South Carolina.**

Below: **Humidor by David W. Dyson.**

TWAIN ON ETIQUETTE

As famous as Mark Twain was with his classic stories and zinging one-liners, he was equally notorious for his constant smoking of cheap cigars. On one of his journeys, he stayed with friends and left cigar ashes all over their home—on the mantle, the piano, and on the window sills. Reverent fans of his work, members of the household quickly scooped the ashes into a mason jar and requested Twain's autograph to prove their authenticity. Twain happily obliged, inscribing on the label, "These are positively my ashes. — S.L. Clemens."

"To cease smoking is the easiest thing I ever did; I ought to know because I have done it a thousand times."

—MARK TWAIN

THE TOBACCO MYSTIQUE

for an audience than by attaching universally recognized actions to characters and demanding a reaction from the reader. By the very fact that smoking is and always has been so potent a social phenomenon, its inclusion in written works inspires an infinite number of similar (based on the universality of smoking), but not identical (because of the reader's individual perceptions of, and experiences with, smoking) character interpretations. A character identified with by a broad spectrum of readers is an archetype. Once an author has established an archetype, he or she has successfully connected with his or her reader.

The fascination propels itself. Mark Twain's mockery of the anti-tobacconists has been built upon tenfold by both his peers and more contemporary writers. *Drinking, Smoking, and*

Above: **The caterpillar smoked from a long hookah in Lewis Carroll's *Alice's Adventures in Wonderland*.**

Left: **Novelist John Steinbeck takes a drag at his home in Sag Harbor, Long Island, in 1962.**

Opposite: **Mark Twain, pictured here with pipe in hand, was known as a lover of cheap cigars.**

THE TOBACCO MYSTIQUE

"I will never consider myself a nonsmoker, because I always find smokers the most interesting people at the table."

—MICHELLE PHEIFFER

Screwing (Chronicle Books, 1994) is a collection of short stories, essays, and poems by past and present writers who revel in the decadence that so offends mainstream society. Henry Miller, Anais Nin, Charles Bukowski, and Fran Lebowitz, "America's #1 chain-smoking, tough-talking headmistress of wit," are but a few of the creative minds featured in this well-needed celebration of how not to behave when trying to appease the masses.

Even the debate about tobacco has become fodder for novels. Christopher Buckley's *Thank You for Smoking* (Random House, 1994) is a satirical look at how the anti-smoking campaign and its resultant effect on a nervous industry threatens to maim the previously tranquil life of one tobacco industry spokesperson.

Smoking In Film

When film makers started including pipes, cigars, and cigarettes in their movies, they set a precedent and an artform from which few movies can entirely extract themselves. A smoking character has an identity even before speaking; a statement is made when tobacco is lit. A movie's smokers use the prop to enhance the personality of their character. The rebellious one, the fashionable one, the sophisticated one, the seductive one, the depraved one; these traits come alive through a certain walk, the movement of an eye, and often, a dose of burning tobacco.

Audrey Hepburn strolls through *Breakfast at Tiffany's* as the hopelessly tasteful scatterbrain, Holly Golightly. The footlong filter dangling from her lips tells us she is quirky from the outset and continues to do so throughout the movie. However, characters like Cruella De Vil, the evil dognapper in Walt Disney's *101 Dalmatians*, smokes from that same type of long black filter to create the portrayal of a sinister and

The paparazzi has a long tradition of capturing dashing Hollywood stars lighting up. *Right:* **One would think that someone from Cuba would prefer a cigar...A young, pre-Lucy Desi Arnez packs a pipe in New York City, 1940.** *Opposite:* **Laurence Fishburn blazes up a stogie at the opening of L.A.'s Grand Havana Room, 1997.**

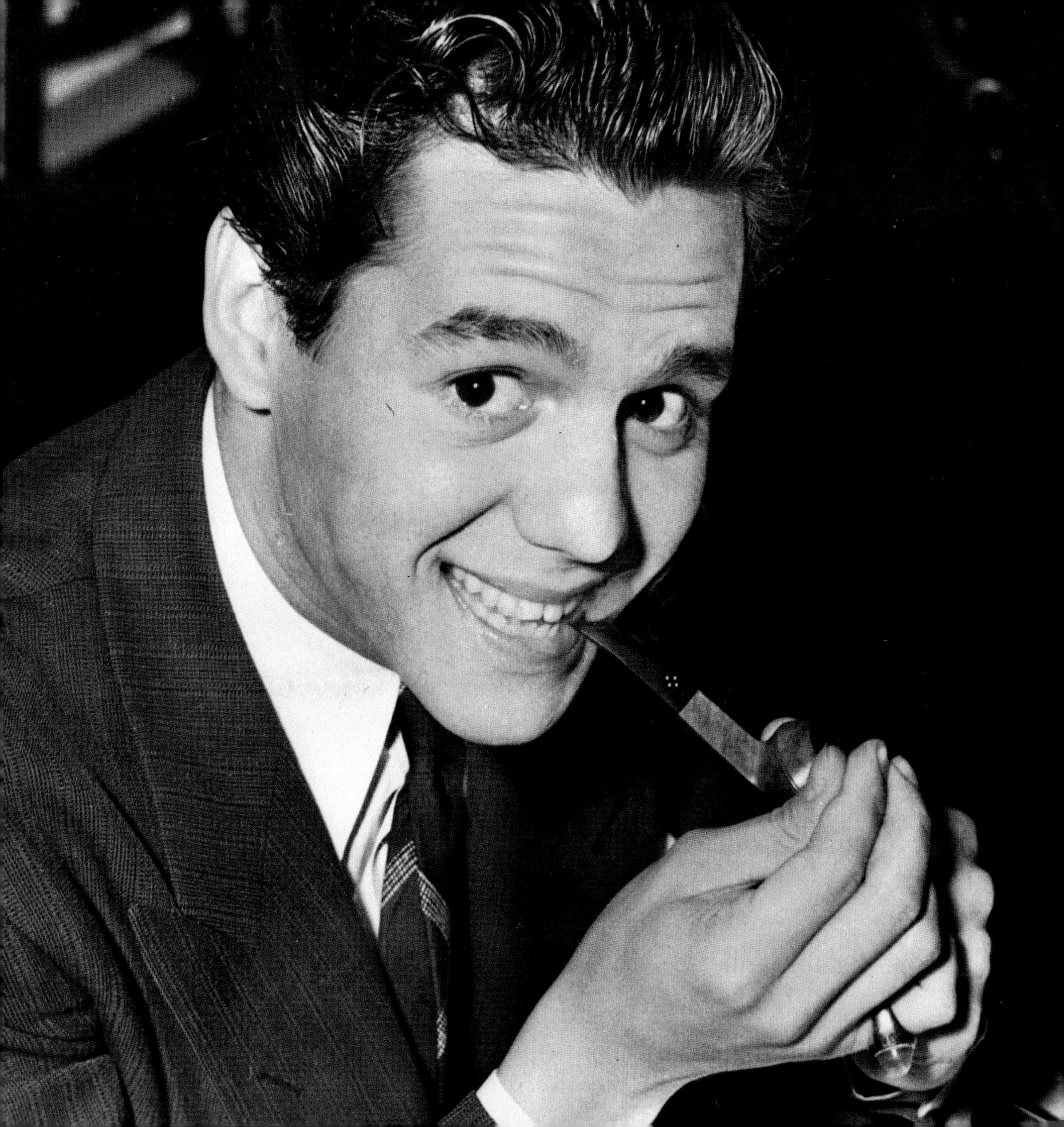

Right: **Gloria Grahame and Glenn Ford blaze up in *Human Desire*, 1954.**

Opposite: Bob Hope gives new meaning to the act of smoking in *Alias Jesse James*, 1959.

THE TOBACCO MYSTIQUE

A VISUAL THRILL AS WELL

Characterization is not the only reason directors incorporate smoke into films. While they demand specific movements from their actors and actresses, it is not always enough to bring their vision for a scene onto the screen. Location, color, lighting, and movement are integral to quality film making. Where an actor's efforts may not quite do the trick, a winding wisp of opaque smoke could provide just the right movement for a scene. A cigarette allows the camera to roll through silence and provides visual stimulus for scenes with otherwise minimal action. Smoke's movement across a screen maintains the beat of a movie in flow much the way a metronome marks the rhythm of a piano piece in play.

frightfully twisted torturess. And who would recognize the cunning Penguin, from the Batman series, without his speech impairing cigarette-and-filter lip medallion. The same prop is used by all of these performers to a drastically different end, but in remarkably memorable ways.

The Smokers Of Early Film

Humphrey Bogart and James Dean sent the image of male smokers soaring. Bogart secured himself a place in the thinking man's tough guy hall of fame, while Dean turned rebellion into Hollywood's hottest commodity. Spencer Tracy, James Cagney, Fred Astaire, and Henry Fonda also created memorable images through the mist of their tobacco. Groucho Marx's ever-present cigar, the pipe of Sherlock Holms, and the hundreds of cigarettes smoked by film's other early leading men were upstaged only by the glamour infused into smoking by film's early female stars.

SMOKE

115

THE TOBACCO MYSTIQUE

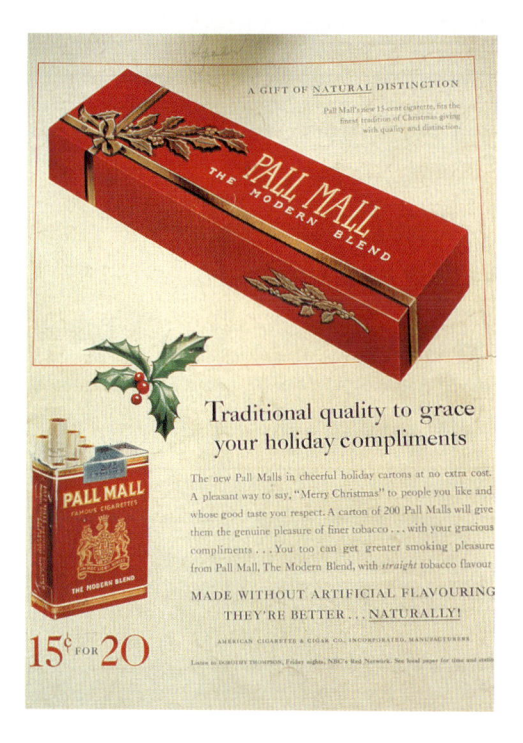

Bette Davis, Lauren Becall, Ginger Rogers, Barbara Stanwyk, and the list goes on: These are the women who helped spawn a nation of female smokers with their larger than life portrayals of tobacco's inherent mystique. Marlene Dietrich smokes in *Seven Sinners*. Lana Turner makes us inhale as she smokes throughout *The Postman Always Rings Twice*. Later, Anne Bancroft smokes in *The Graduate*, and the list goes on and on. Elizabeth Taylor, Sophia Loren, and Natalie Wood took up tobacco in their films and whether the goal was sensuality, raw defiance to social mores, or an unnamed middle ground, the effects were staggering to audiences yet unjaded.

Contemporary Classics

While modern-day filmmakers can never revive the newness and shock value that smoking sparked forty years ago, it's unlikely that they would give up this versatile prop. Just as authors use smoking to incite familiarity or even bias in their readers, directors can

Opposite: A fabulous cigarette meets with a less than fabulous hat: Audrey Hepburn's Holly Golightly gets the party—and the fire—going in *Breakfast at Tiffany's*, 1961.

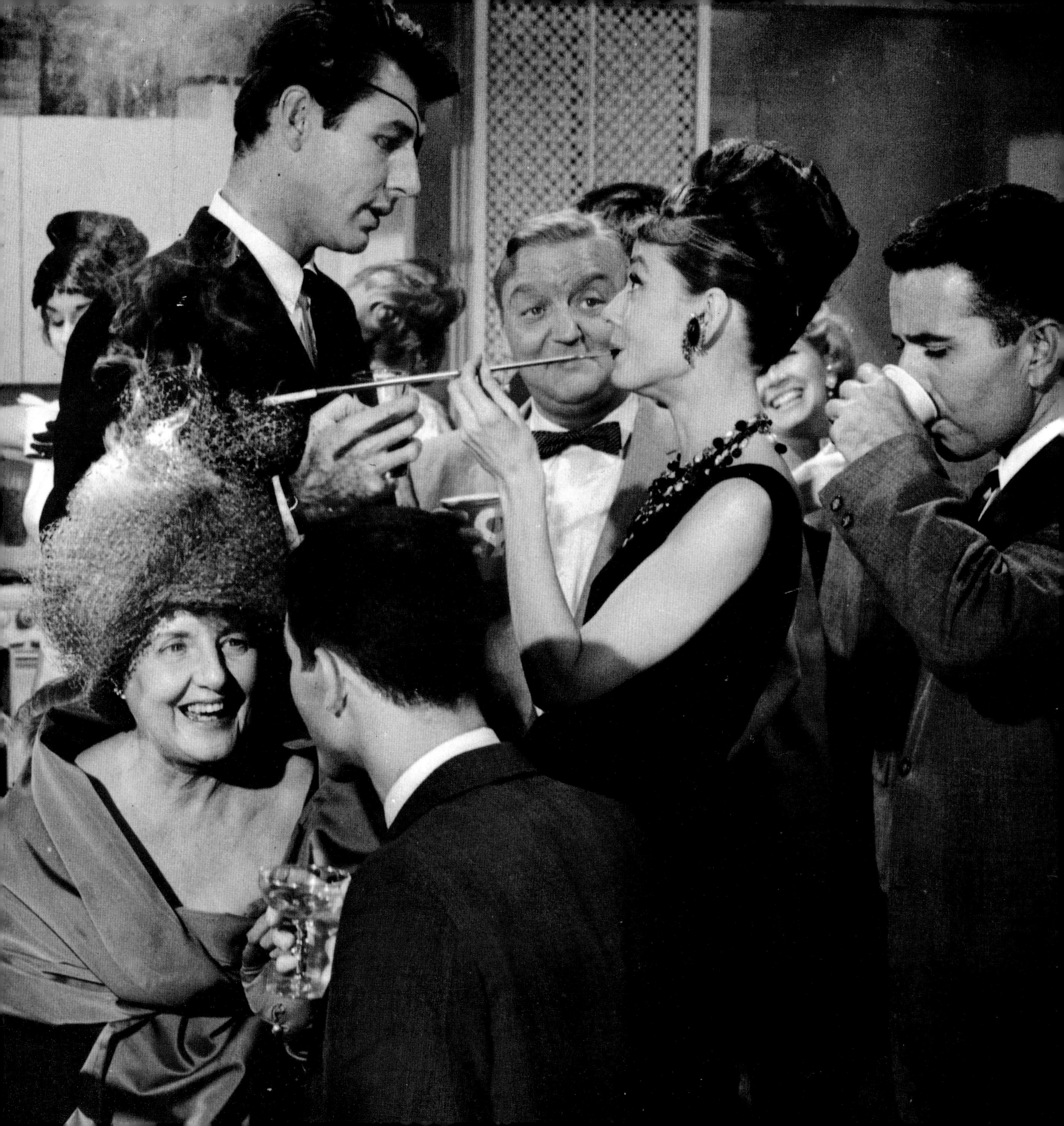

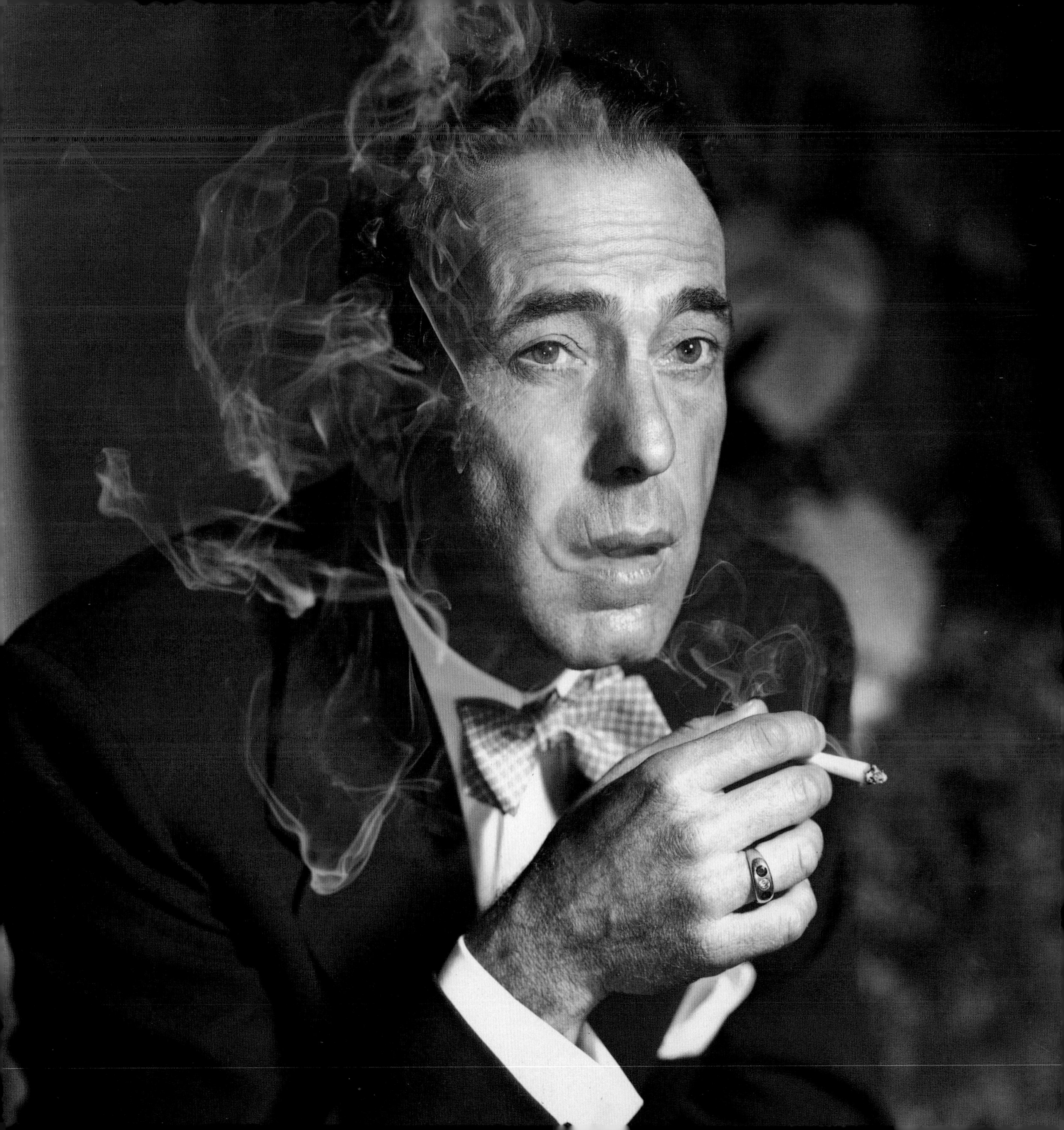

THE TOBACCO MYSTIQUE

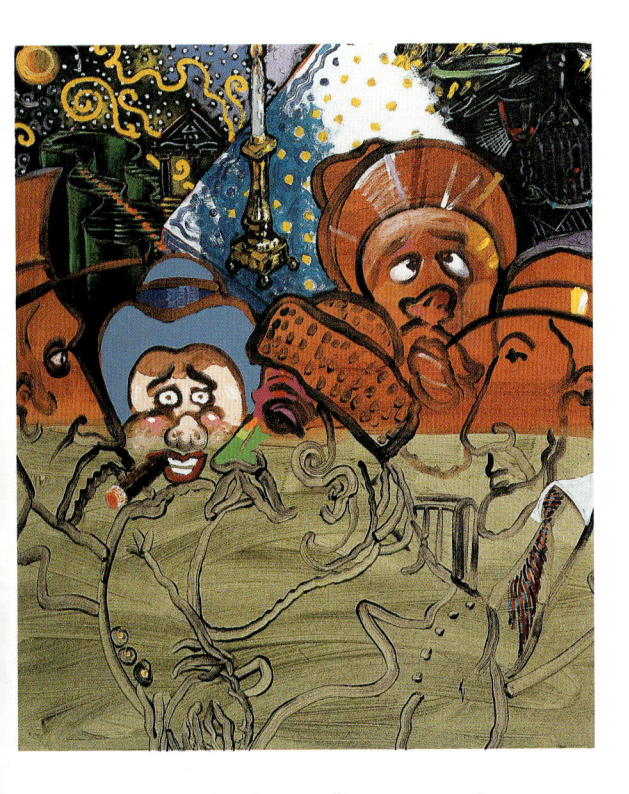

now manipulate tobacco to stir up any number of positive or negative images we've come to associate with it. The intentions of Oskar Schindler in Steven Spielberg's 1993 blockbuster *Schindler's List* are hidden through a veil of smoke; in Scott Hick's brilliant film *Shine*, Australian pianist David Helfgott's eccentricities are captured in the manic grasp of his cigarette. James Bond has gone back and forth between smoking cigarettes and/or cigars throughout many of his castovers.

There has been no slowing down in film's female smokers, either. Barbara Streisand, Diane Keaton, and Meryl Streep have used the image in dozens of ways. The familiar faces of our most recent generation of actresses who've smoked on camera include: Winona Ryder, Julia Roberts, Lili Taylor, Jennifer Jason-Leigh, and Uma Thurman. In the gray area between actor and character role, many performers sharpen the edges of their created persona with props, and many directors make tobacco the prop of choice.

It is truly a fascinating accomplishment on the part of media's creators that something so small as a cigarette, cigar, or pipe could sway perceptions so staggeringly. While many directors are still keen to utilize tobacco and smoking to any advantage they can, some have gone the riskier route of deglorifying its past media mysticism. In today's strongly anti-smoke society, screenwriters have found yet another stepping stone to public conscience. In Kenneth Branaugh's *Dead Again*, Andy Garcia plays a nicotine-charged newspaper reporter who chain-smokes throughout

TOBACCO IN ART

Jackson Pollock (1912-1956), an American artist and forerunner of abstract expressionism, created some of his greatest pieces by dripping paints and enamels onto huge canvases stretched out across the floor. He is considered by many to be the quintessential action painter. Oftentimes, as he was painting, his cigarette ash or butt would fall from above onto the canvas— Pollack would leave it there to dry within the intricate lace of his creation.

Opposite: When Humphrey Bogart was hidden beneath a haze of smoke, it was hard to figure whether he ws playing a good guy or a bad guy.

Left: The Rabbis II, by Archie Rand, 1985.

Right: Native Americans gave the rest of the world tobacco, and the rest of the world never forgot. Cigar store Indian, late nineteenth century.

SMOKE

"Smoking is, if not my life, at least my hobby. I love to smoke. Smoking is fun. Smoking is cool. Smoking is, as far as I'm concerned, the entire point of being an adult."

—Fran Lebowitz

A lady never lights her own ... Paul Henried offers Bette Davis a light in *Now, Voyager*, 1942.

SMOKE

THE TOBACCO MYSTIQUE

his career. When Branaugh finally shows us Garcia as an old man, he is hospital ridden and smoking through the tracheotomy tube in his throat.

In the great debate that is tobacco, lies yet another path to public cognizance; and the attention of the public is income for the industry. For the film industry, this turning leaf of negative press for tobacco may well keep the debate heated and, therefore, the topic appealing to masses of moviegoers.

A Precarious Future?

Perhaps the pop culture pendulum's recent swing back toward cigars will render the cigarette obsolete. It's always been the haut monde that has led the way for each of tobacco's numerous manifestations and with such famous stars as Bill Cosby, Shari Belafonte, Arnold Schwarzenegger, Ted Danson, and Demi Moore puffing on the finer end of the tobacco train, we can be sure of heightened attention in that direction. Cigars, though, are a

Right: **Harvey Keitel is selling, and Stockard Channing is buying.** *Smoke,* **1995.**

Above: **Colibri's Churchill V combination cigar cutter and lighter.**

SMOKE

123

THE SOUND OF SMOKE

From the deadly-hip jazz clubs of the 1930s, to the vibrating warehouses of the modern international rave scene, marijuana (a Mexican word for cannabis) has cut deep grooves into music's illustrious tracks. Louis Armstrong hailed its virtues before most knew what it was, Bob Marley and Peter Tosh openly praised it as a sacred herb, the Beatles, the Rolling Stones, and, of course, Bob Dylan all went public with their penchant for pot.

Slicing through traditional barriers, cannabis and music teamed up in the 1960s to spearhead a new social awareness and a more peaceful approach to coexistence. Proponents of legalized marijuana use spoke of pot's mellowing effects and enhanced creativity. Never before had a generation so wholly embraced what mainstream society considered shocking, licentious, and dangerous; and never before had two such simple pastimes as music and smoke earned the attention that they did during this time.

The music industry remains inextricably bound to its rich history of experimentation and controversy while the battle for decriminalization of marijuana moves into larger arenas. Whatever their bond today, few can dispute that music and musicians have played a great role in pot's past and many will agree that pot has contributed as much to music's own ancestry.

Every Hollywood tough guy's got to have one: Baby Herman chomps down on a stogie in Who Framed Roger Rabbit?, **1988.**

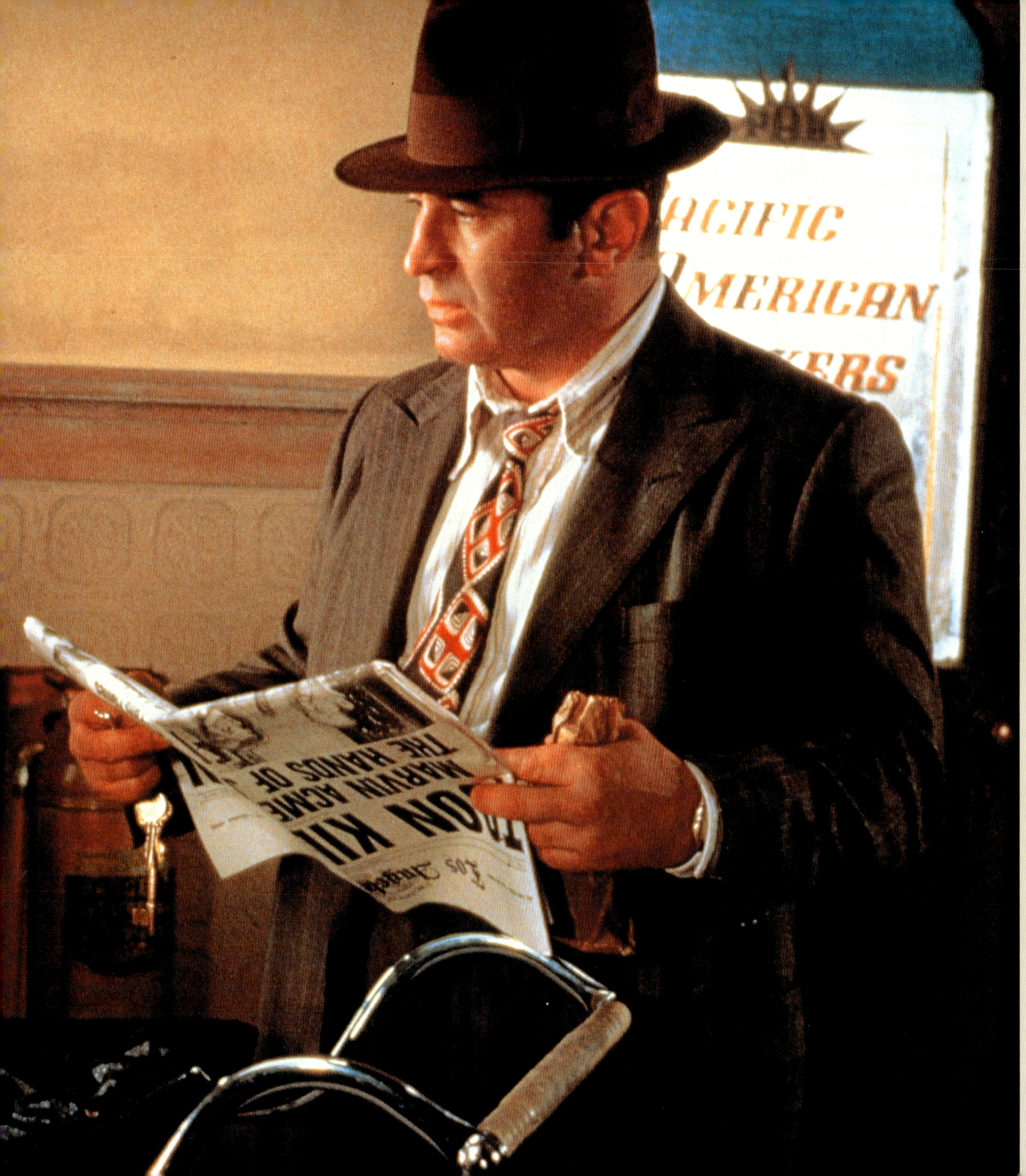
MARVIN ACME
THE HANDS OF

THE TOBACCO MYSTIQUE

"I'm glad I do not have to explain to a man from Mars why each day I set fire to a dozen little pieces of paper, and then put them in my mouth."

—MICHAEL MCLAUGHLIN

Left: **Smoking is glamorous again, but now it's not cigarettes: Rosanna Arquette puffs away on a ciger, 1996.**

Right: **Did crooner Bing Crosby use a pipe to keep his own pipes in tune?**

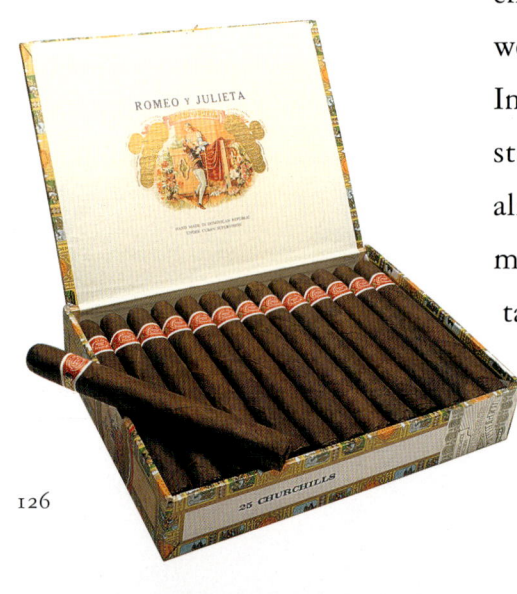

small part of tobacco's playing field and it is the entire industry whose future remains in waiting.

No one could have foreseen how deeply entrenched tobacco would become in our world. Who then, can name its future? In societies partial to trends and wary of stagnation, this mystical herb has already made an impressive mark on the moving target that is our collective tastes. More than any one definitive reason, the appeal seems to be a combination of tradition with modern hype, statement with silence, and vice with value.

As government regulation and social pressure thwart the once unstoppable growth of the tobacco giant, many predict its unequivocal extinction in the next few years. Others are floored by the rate at which it is moving into second and third world countries and argue that tobacco hasn't even approached the height of its use. Certainly, the only true test is the passage of time, for ultimately, tobacco use is a personal choice and despite the power of advertising, media, government, and opposition movements, our greater fascinations lie in our individuality and our right to do what is best (or even worst) for ourselves.

INDEX

Pages numbers in boldface indicate photo captions.

advertising:
for European theater, 91
for Job cigarettes, 11, **102**
for Marshall's cigarettes, 87
for Prince Albert pipe tobacco, 48
on television, 91, 101–3
women in, 88
air curing, 48
Allen, Gracie, 62–63
American Machine & Foundry Co., 6
Arawak Indians, 16
Arents, George, Jr., 6–7
Armstrong, Louis, 124
Arnez, Desi, 113
Arnold's Tobacco Shop (NYC), 80
Arquette, Rosanna, 126
Astaire, Fred, 91, **115**
Auerbach, Red, 79

Baby Herman, in *Who Framed Roger Rabbit?*, 124–25
Bacall, Lauren, **116**
Bahari, *Rajah of Chambra*, 39
Ball, Lucille, in *Lover Come Back*, 81
bamboo pipe smoker, 40
Bancroft, Anne, in *The Graduate*, **116**
Batman series, **115**
Beatles, the, 124
Belefonte, Shari, 123
betel nut, 10
bhang, 10
blending, 54
Bogart, Humphrey, **115**, 118
Bordoni, Irene, 88
Briar Pipes, 44–47
Brooks, Jerome, 6
Brosnan, Pierce, in *Mars Attacks!*, 52–53
Buckley, Christopher, *Thank You for Smoking*, 112
Bukowski, Charles, 112
Burley leaf, 32
Burns, George, 9, 62–63

Cage, Nicolas, 65
Cagney, James, **115**
Camel cigarettes, 87, 101, **104**
casings, 54
Castro, Fidel, 66
caterpillar smoking (*Alice's Adventures in Wonderland*), 111
Cézanne, Paul, *The Pipe Smoker*, 35
Channing, Stockard, in *Smoke*, 122–23
chewing tobacco, 31, 32
children's toys, tobacco-related, 54
Churchill, Winston, 17
cigar bands, 76–79, **84**
cigar boxes, 76, 84
labels of, 72, **73**
cigarette card ladies, 88, **109**
cigarette holder, pretzel, 88
cigarette packages, 87
cigarette rolling machine, 87
cigarettes, 29, 30–31, 32, 80, 82–104
affectations and tricks with, 98
business of, 31
dangers of, 98–103
machine- vs. hand-rolled, 31, 85
cigar factory workers, 70
"cigar nights," 79
cigar rollers, 64
cigars, 17, 27, 29–30, 58–80, **58**, 62
resurgence of, 79–80, 123–26
silent pleasure of, 75
cigar store Indian, 119
Cloud, Hugo, collage, 71
clove cigarettes (kreteks), 95
Colbert, Claudette, 100
Colibri's Churchill V combination cigar cutter and lighter, 123
collectors, 84
Columbus, Christopher, 16
Connecticut, 62, 71–72
Cooper, Gary, 83
Cosby, Bill, 123

Crosby, Bing, 127
Cruella de Vil, 112–15
Cuba, 61–67
Cuba Club (Miami), 76
Curtis, Tony, 91

Danson, Ted, 123
Davidoff, Zino, 7
Davis, Bette, **116**
  in *Now, Voyager*, 120–21
Davis, Sammy, Jr., 96–97
Dead Again (film), 119–23
Dean, James, **115**
Depp, Johnny, 91
Dickson, Sarah A., 6
Dietrich, Marlene, 5, 91
  in *Seven Sinners*, **116**
dragon lighter, 9
Drinking, Smoking, & Screwing (fiction anthology), 111–12
Duke, James, 85–87
Dunhill, Alfred H., 7
  The Gentle Art of Smoking, 34
Dylan, Bob, 124

earth pipes, 39
European explorers, 12, 13
Evangelista, Linda, 64

filtered cigarettes, 103
Fishburn, Laurence, 112
flavorings, 28–29, 32, 47, 54
flue-curing, 48–51
Flynn, Errol, 8
Fonda, Henry, 91, **115**
Ford, Glenn, in *Human Desire*, 114

Gabor, Eva, 91
Grahame, Gloria, in *Human Desire*, 114
greeting card, 54
guillotine cigar cutters, 16, **78**

Hacker, Richard Carleton, *The Ultimate Pipe Book*, 7
Halbert, Jacques, *Cigar Landscape*, 61
hashish, 104
Havana cigars, 76
heather wood, 43–44
Heimann, Robert K., *Tobacco and Americans*, 32
hemp, 43
Henried, Paul, in *Now Voyager*, 120–21
Hepburn, Audrey, in *Breakfast at Tiffany's*, 112, **116–17**
High Times magazine, 104
Hogarth, William, *Election Entertainment*, 21
Holden, William, 91
Hollywood glamour photo, 89
Homer, Winslow, *The Briarwood Pipe*, 24
Hope, Bob, 91
  in *Alias Jesse James*, **115**
Hudson, Rock, 91
humidors, 56, **109**
H. Upmann Petit Coronas, 67

James Bond, 119
James I, King of England, 27–28
  Counterblast to Tobacco, 27
Jason-Leigh, Jennifer, 119
Java, 72
Job cigarettes, 11, **102**
Joe Camel, 101

Keaton, Diane, 119
Keitel, Harvey, in *Smoke*, 122–23
Kennedy, John F., 67, 107
Klein, Robert, *Cigarettes are Sublime*, 109
Kool Cigarettes, 95

Laurel and Hardy, 91
Lebowitz, Fran, 112
Leino, Clarence, *You're Welcome!*, 46
Letterman, David, 68–69
Lewine, Harris, 98
Loren, Sophia, **116**
Lucky Strike, 91, 98
lung cancer, 101

MacArthur, General Douglas, 98
Madonna, 68–69
Manet, Edouard, *Portrait of Stephane Mallarme*, 74–75
Marie Antoinette, scene from, 16
marijuana (cannabis), 104, 124
Marlboro cigarettes, 103
Marley, Bob, 124
Martin, Dean, 96–97
Marx, Groucho, 60, **115**
Masulipatam, 29
matches, 47, **56**
Maugham, W. Somerset, 55
McCarthy, Cormac, 109
McCarty, Kim, *You Talkin' to Me?*, 33
meerschaum pipes, 21, 47
Mermaid Pub (London), 44
Mild Seven cigarettes (Japan), 88
Miller, Henry, 88, 112
Moore, Demi, 123
  on *Cigar Aficionado* magazine, 59
Munch, Edvard, *Self Portrait with Cigarette*, 96

Native Americans, 7, 9, 13, 14, **39**
Nicholson, Jack, 105
Nicot, Jacques, 18
Nicotiana Tabacum, 18, 47
Nin, Anais, 112

opium, 38, 40
opium smokers, 36–37

Palmer, Lilli, 91
peace pipe (calumet), 14, **15**, 39
Peck, Gregory, 91
Pennsylvania, 72
Phillip Morris, 104
Pierre-Renoir, Auguste, *Portrait of Claude Monet*, 20
pipe cleaners, 56
pipe reamers, 56
pipes, 17, 21–27, 28, 31, 34–56, 49, 54–56, 61
  old clay, 39
  religious/medicinal uses of, 39
pipe tobacco, Aromatic and English, 54
pipe tomahawks, Blackfoot Indian, 14
Player Tobacco Company, 92
Pollock, Jackson, 119
puro, 66

radio, 95
Raleigh, Sir Walter, 21–24, **22–23**, 44
Rand, Archie, *The Rabbis II*, 119
Rathbone, Basil, 43
Reagan, Ronald, 91
restaurants and bars, cigar- friendly, 79
Richards, Keith, 91
Richards, Michael, 41
RJR Nabisco, 104
Roberts, Julia, 119
Robinson, Edward G., 45
rock and roll, 91
Rockwell, Norman, *Triple Self-Portrait*, 50
Rogers, Ginger, **116**
Rolfe, John, 24
Rolling Stones, the, 124
Roosevelt, Franklin D., 10
Roth, Tim, 95
Russell, Rosalind, 86
Ryder, Winona, 119

Salinger, Pierre, 67
Sand, George, 109
Sarandon, Susan, 99
Sargent, John Singer, *Asher Wertheimer*, 106
Saturday Evening Post, 87
Schindler's List (film), 119
Schwarzenegger, Arnold, 67, 123
Shaw, Thurstan, *Early Smoking Pipes in Africa, Europe, and America*, 37
Sherlock Holmes, 25, 43, 54, **115**
Shine (film), 119
Shiseido, 88
Shumann-Heink, Madam, 27
silks, 92–95

Sinatra, Frank, 96–97
smoking
  bans on, 27
  in fiction, 109–12
  in film, 95, 112–23
smoking jacket, Savinelli chalkstripe, 71
smoking rooms (divans), 61
snuff, 27, 28–29, 31, 61
snuff boxes, 28, **28**
social pressure, 126
soldiers, and cigarettes, 29, 92–93, 98
Stanwyck, Barbara, **116**
Steinbeck, John, 111
stem-socket pipes, 37
Streep, Meryl, 119
Streisand, Barbara, 119
suffrage movement, 91
Sumatra, 72
Swanson, Gloria, 94
Sweet Caporal cigarettes, 88
sweeteners, 32, 54

Taj Mahal, 7
tampers, 54–56
Taylor, Elizabeth, **116**
Taylor, Lili, 119
television, 91, 101–3
Thurman, Uma, 119
  in *Pulp Fiction*, 6
tobacco:
  aging of, 51–54, **51**
  cultivated, in America, 24
  curing of, 47–51, 108
  as currency, 14
  economic potential of, 14, 17, 30–31
  geographic origins of, 13
  grading of, 51
  history of, 10–32
  modern misuse of, 9
  mystique of, 106–26
  mythic origin of, 17
  opponents of, 21, 27–28, 31, 54, 119–23, 126
  poisonous aspects of, 18
  purported powers of, 18, 22
  sacred functions of, 14
  social functions of, 16–17, 18–21, 24–27
  ubiquitousness of, 4–9
  use in America, 31–32
  Western awareness of, 16–18
Tobacco Institute Research Committee, 101
tobacco pouches, 56
tobacco seal, Cuban government, 76
tobacco tins, 21
torcedors, 64, 66–71
Tosh, Peter, 124
Tracy, Spencer, **115**
Travolta, John, in *Pulp Fiction*, 84–85
Turner, Lana, in *The Postman Always Rings Twice*, **116**
Twain, Mark, 32, 110, **110**, 111

Ventegodt, Niels, 84
Virginia, 21, 27, 32
Vuelta Abajo (Cuba), 66

water pipe (dakka), 40
Webb, Jack, 91
Wheeler, Herbert, 90
White Burley, 32
Willie the Penguin, 95
women, 7, 26, 27, 31, 32, 61, 79, 88
  cigarette marketing to, 91–95
  cigar rollers, 64
  punished for smoking, 27
  smoking on the sly, 26
  and snuff, 28
  torcedors, 71
Wood, Natalie, **116**
Wyman, Jane, 91

128